The Florida
Scrub Jay

MONOGRAPHS IN POPULATION BIOLOGY

EDITED BY ROBERT M. MAY

The Florida
Scrub Jay

DEMOGRAPHY OF A
COOPERATIVE-BREEDING
BIRD

GLEN E. WOOLFENDEN

AND

JOHN W. FITZPATRICK

PRINCETON UNIVERSITY PRESS

PRINCETON, NEW JERSEY

1984

Library of Congress Cataloging in Publication Data will be
found on the last printed page of this book
ISBN cloth 0–691–08366–5   paper 0–691–08367–3
This book has been composed in Linotron Baskerville
Clothbound editions of Princeton University Press books are printed
on acid-free paper, and binding materials are chosen for strength and
durability.
Paperbacks, although satisfactory for personal collections,
are not usually suitable for library rebinding
Printed in the United States of America by
Princeton University Press, Princeton, New Jersey

To the Memory of Richard Archbold,
   explorer, benefactor, and friend

Preface

In 1969, while measuring breeding bird populations in several habitats at the Archbold Biological Station, Woolfenden found eight Florida Scrub Jay nests, six of which were attended by more than two adult-plumaged jays. The jays were relatively tame and were easy to color-ring for individual recognition. By 1975, these jays had worked their way to a centerfold illustration in E. O. Wilson's monumental synthesis, *Sociobiology*. They arrived there by being near the center of a growing controversy about how cooperation and "altruism" evolve within animal populations. Fully mature Florida Scrub Jays often help to raise offspring that are not their own, forming a classic case of "helping-at-the nest" and apparently providing a lucid example of altruism in nature. For a Florida Scrub Jay, helping to raise extra sibs would seem to be nearly, or exactly, as valuable to an individual as raising its own offspring. Wilson (1975, p. 454) cited the jays, among other cooperative bird species, as probable examples of kin selection at work.

At the same time, many investigators had begun questioning the very existence of so-called altruistic acts in nature. In large part this became a semantic discussion, and in general the ambiguous term *altruism* has been mercifully dropped from sociobiological jargon. However, the more interesting part of this controversy continues. We still remain uncertain about the exact role played by kin selection in the evolution of helping behavior in general, and cooperative breeding in particular. Theoretical models clearly support the potential importance of kinship in advanced animal sociality. Shared genes of next-generation kin are as valuable to an individual as are the same genes in its own offspring. Yet, certain social features in many wild populations have evolved independent of kinship. Therefore, kin selection might be a red herring amidst social systems that have evolved purely for other reasons. Field tests to sort out such tricky questions, especially with birds, remain difficult and time consuming, although this has not curbed the tide of writing about this issue. One needs only to peruse the proceedings of the 1980 Dahlem Conference (Markl, ed., 1980) to appreciate how slowly the voluminous field of sociobiology has progressed in answering some of the truly fundamental issues raised first by Hamilton (1964), then more provocatively by Wilson (1975).

Long-term demographic studies of natural populations are, by all standards, among the foremost methods for sorting out these evolutionary issues. As it turned out, Florida Scrub Jays were uniquely suited for such a study (see chapter two), especially in that pairs both with and without helpers always exist within the population. This allows us to ask questions about whom, when, and how helpers help, using built-in controls. All these, in turn, lead to the most important evolutionary question: why do helpers help? By the end of our first decade of field work, our data strongly suggested to us that some popular and potentially far-reaching conclusions regarding the evolution of cooperation might have been more attractive than accurate. A social system that seems born out of a spirit of familial generosity instead emerges as far more selfish and individualistic than early sociobiologists suspected.

In 1979 we thought it time to summarize these findings. With the financial backing of the National Geographic Society and a Frank M. Chapman Fellowship of the American Museum of Natural History, the two of us spent nearly four months together in an intensive session of data analysis and preliminary writing (atop the annual work load of a Scrub Jay breeding season). This book is a result of those efforts, drawn together during subsequent, periodic bouts of writing and follow-up analyses that were squeezed into the brief periods when we were not 1,500 km apart or on different continents.

The decade of field work summarized herein includes the labors of many. It is our great pleasure to acknowledge the help of some of these colleagues. Four students have completed Master's Degrees on Florida Scrub Jay topics, and in doing so G. Thomas Bancroft, D. Bruce Barbour, Anthony R. DeGange, and Jerre A. Stallcup contributed importantly to the project as a whole. Among the many other students who have helped, we especially thank Brian Harrington, Carole Hendry, Debra Moskovits, and Ralph Schreiber. Debra Moskovits also served as an outstanding long-term field assistant. Two other persons have volunteered their long-term service in the field: Bobbie Kittleson, our provider and faithful critic, and Chester Winegarner, friend and devoted protector of the jays. Recently, Jan Woolfenden has put in much helpful field time. We most sincerely thank all these contributors.

Our conversations about jays and cooperative breeding have been rich and extensive with colleagues too numerous to list in detail. In this respect, we especially thank Russ Balda, George Barrowclough, Jerry and Esther Brown, Steve Emlen, Jack P. Hailman, Wayne Hoffman, Henry Horn, Richard Kiltie, Walt Koenig, John Krebs, Marcy Lawton, Cliff Lemen, Robert May, Earl McCoy, Ron Mumme, Frank Pitelka, Ulli Reyer, Jeff Walters, and Amotz Zahavi.

Throughout the project, the Archbold Biological Station and its staff have provided extensive logistical support. We are deeply indebted to the late Richard Archbold, founder, and to James N. Layne, Director. The National Geographic Society provided the funds needed for our initial period of analysis and writing and for our field assistant in 1979. A Chapman Fellowship to Woolfenden was a boon to getting the job done. Woolfenden also has received support from the American Philosophical Society, the St. Petersburg Audubon Society, and Mr. and Mrs. Lewis Wiederhold. Fitzpatrick received additional support from the Conover Fund of the Field Museum of Natural History, Chicago, and from research funds of the Department of Biology at Princeton University.

During the long-awaited final push, Doris Nitecki, Dianne Maurer, and several members of the secretarial staff in the Department of Biology, University of South Florida, diligently typed (too many) drafts of text and tables. We thank them for their long efforts. William E. Southern carefully read page proofs, and we thank him sincerely.

The two of us derive great satisfaction from this continuing, cooperative undertaking, and we accept all blame for its shortcomings. This book has been a truly joint venture, from field work to data analysis to writing. Its order of authorship was determined by a coin toss.

John W. Fitzpatrick   Glen E. Woolfenden
Chicago, Illinois   Tampa, Florida

Contents

CONTENTS

The Florida
Scrub Jay

CHAPTER ONE

Introduction

1.1. THE PROBLEM

Just past dawn on a spring morning in central Florida, one can witness a sequence of events that occurs virtually nowhere else in North America. We introduce the subject of our study with a brief description of one such scenario: After silently leaving her nest with a short, quick flight, a female jay begins bathing and preening in the dense, dew-soaked leaves of the oak scrub. She warms for a few minutes in the early morning sun, forages briefly for insects near the ground, then returns to her nest. Three naked nestlings with mouths wide open crane skyward upon her arrival. As she perches over them on the nest rim, the female is joined by *two* male jays, each with insects in its bill. One, with a quiet guttural call, passes the food to the female, who promptly transfers it to a nestling. The other provider perches briefly on the nest rim and feeds a second nestling directly. Both males soon fly off. As the female settles back upon the nestlings, a *fourth* jay, this one a female, flies into a shrub a meter or so away, watching. A typical day of nest activity has begun (see frontispiece).

The birds just described are Florida Scrub Jays (*Aphelocoma c. coerulescens*), and the interactions among the four jays have been variously termed "helping-at-the-nest" (Skutch, 1935, 1961), "group" or "communal breeding" (Zahavi, 1976; Brown, 1978a) or, as we prefer, "cooperative breeding" (Lack, 1968; Brown, 1970; Rowley, 1976; Emlen,

1978; Fry, 1977). This social system is characterized by the participation of more than a mated pair in the production of young. As stated in a recent, comprehensive review by Brown (1978a, p. 124), "Communal breeding as a social system is defined primarily by the regular involvement of *helpers* in the feeding and care of the young" (italics ours), where a helper is defined operationally as "a bird which assists in the nesting of an individual other than its mate" (Skutch, 1961). The pages to follow provide a comprehensive analysis of one such social system, and the demographic patterns that accompany it, in a wild, free-living population of birds.

The evolution of cooperation among animals has fascinated biologists at least since Darwin (1859), who mentioned it as a seeming paradox in his theory. The topic has received considerable attention ever since Hamilton (1964) and Maynard Smith (1964) pointed out the plausibility of the idea that aid giving arose through kin selection. In theory at least, individuals that sacrifice their own breeding efforts in order to help (i.e., raise the fitness of) others sometimes could be selectively favored in the long run if the donor and recipient of such aid share sufficient genetic relationship as kin. Kin selection has been widely, and sometimes uncritically, cited as an essential process underlying the evolution of cooperative breeding in many birds (e.g., Maynard Smith and Ridpath, 1972; Wilson, 1975; Brown, 1974, 1978a; Watts and Stokes, 1971, but see Balph et al., 1980).

Because an excellent recent review of avian communal breeding systems is available (Brown, 1978a), we do not attempt one here. However, we emphasize a few of the points made by Brown in his review (see Fitzpatrick and Woolfenden, 1981).

4

1. Because cooperative-breeding birds are diurnal and relatively easy to observe, they probably are the best vertebrates to study when looking for test cases regarding the operation of kin selection under natural conditions. Features of their helping behavior and its effects can be measured with precision; successive generations of individually marked offspring can be followed from birth to death; finally, independent genealogical records can be monitored in a single, long-term study because home ranges are relatively small. To Brown's (op. cit., p. 123) list of advantages we add that in many bird species, including the Florida Scrub Jay, natural controls are provided for a number of critical tests if a population is studied over many years. For example, the effects of helpers can be examined directly (see chapter eight), and we can postulate realistic demographic characteristics for "ancestral" populations in which helping was just evolving (see chapter ten).

2. Controversy still exists regarding the role and importance of kin selection in the evolution of avian cooperative breeding (e.g., extensive discussions in Markl, 1980). Because adequate field data have been lacking until very recently, this controversy is based on theoretical considerations, and even opinion, more than on detailed interpretation of real natural systems. Opinions range from the extreme that kin selection does not operate at all in natural populations (Zahavi, 1974, 1976, personal communication) to the opposite extreme that kin selection is often a necessary ingredient in the selective regime leading to helping (e.g., Brown, 1974, 1978a). We have been cited as proponents of the former extreme (e.g., by Brown, 1978a,b). Our position has always been that whereas inclusive fitness is perforce raised if helpers help produce additional kin the relative magnitudes of this increase compared to the many

other, purely individual fitness benefits to the helper *have never been analyzed* (Woolfenden and Fitzpatrick, 1978a,b). We are aware that kin selection *could* lead to helping behavior, but it remains to be shown in even one natural system that kin selection was necessary for helping to evolve. In short, we emphasize that the *relative* importance of the direct versus the indirect components to inclusive fitness still is unclear (terminology used by Brown, 1980). This question requires extensive data before being addressed. It forms one of the themes of this book.

3. All students of avian cooperative breeding agree that absence of the necessary data has prevented proper evaluation of the role kin selection plays in natural populations. Perhaps the question is indeed intractable (Wade, 1979, 1980a,b), but Brown (1978a, 1980) has identified many of the necessary measurements any such evaluation would require. Additional ones will be identified and examined in this treatise. Critical variables include: age- and sex-specific survivorship and dispersal characteristics among non-breeders and breeders; fecundity schedules under a variety of breeding circumstances; effects of helpers on reproduction; the life stage at which these effects are expressed; and measures of the between-year variance in these demographic measures. A glance through any advanced ecology text or pertinent literature review (e.g., Ricklefs, 1979; May, 1981) shows that exhaustive studies of natural demographic patterns outside the laboratory are rare for animals in general, and entirely lacking for cooperative-breeding birds. Disagreements about how helping behavior arises cannot be resolved until such information is available, so that realistic selection models can be built. This infor-

mation can be obtained only through long-term studies of single populations of individually marked animals (Fitzpatrick and Woolfenden, 1981).

1.2. OUR APPROACH

This book is primarily a compendium of demographic information gathered on a natural population of individually color-ringed Florida Scrub Jays. The study was begun by Woolfenden in the spring of 1969. Since that time we have continually monitored the entire jay population—its breeding and territorial activities and the births, lives, movements, and deaths of its individuals—in about 400 ha of oak scrub in south central Florida. We include in this report data accumulated through 1979, thus representing the first decade of the ongoing study.

Florida Scrub Jays live on permanent, 4- to 15-ha territories. About half of these territories at any one time have nonbreeding, adult-plumaged helpers living with the breeding pairs. Therefore, in part, this is a study of cooperative breeding, its effects, and its evolution in a bird population. We hope that the hypotheses and conclusions we develop will add to the growing general framework of social theory.

In addition, we believe that documentation of the basic demography of *any* wild population of animals is itself of paramount importance in ecology, if only because such case studies still are rare. Therefore, we present and summarize, as systematically as we can, data pertaining to all phases of individual and population life histories. Many uses and interpretations of these data are included in our analyses

7

where pertinent to the question of cooperative breeding. We trust that additional uses will be discovered by demographers and other ecologists.

We agreed early in the study that following the natural course of events in the population would be our first field priority. Therefore, we have engaged in no experimental manipulations involving either removal of individuals or changing of habitat characteristics within the study tract. Several studies have included such manipulations in order to examine specific questions about the effects of helpers or habitat structure in a cooperative-breeding system (e.g., Brown et al., 1982; Brown and Brown, 1981b; Brown and Balda, 1977). For our purposes, most such "experiments" are performed naturally with sufficient frequency to provide the necessary data (along with all the necessary background information) to address these questions. Some small-scale behavioral experimentation is possible (and has been performed) because many of the birds we follow are tame (see chapter two).

1.3. CASE HISTORIES

Throughout the text we have inserted consecutively numbered case histories to illustrate certain points or to describe events that provide exceptions to certain patterns. For the reader's convenience, these passages are indented. Although they may be skipped without losing the meaning of the main text, we believe they add significant insight into the social biology of the birds. We present these cases in an abbreviated format but still include all details we consider pertinent. Usually we give the color-ring names of the key individuals (see chapter two and Woolfenden, 1974)

8

to allow the interested reader to relate any one case history to others presented in this and other publications on the marked jays of the Archbold Biological Station.

1.4. ORGANIZATION

All facets of the life history of any animal are so interconnected that any linear sequence of data presentation in a work such as this must be somewhat arbitrary, and at times frustrating. Data regarding any one feature of the Florida Scrub Jay social system bear directly on other features presented and discussed elsewhere. To minimize this problem we have liberally cross-referenced tables, figures, and discussions throughout the text. A synopsis of the important factual results precedes most chapters. We also present summarized conclusions at the end of each chapter, in which we offer our views of how the findings in that chapter bear on those elsewhere in our study and in related literature.

Following a description of our field and analysis *procedures* in chapter two, we begin the story (chapter three) with an introduction to the current and historical *ecological setting* within which Florida Scrub Jays survive. A description of Scrub Jays in western North America and our data on habitat use and population densities in Florida appear in this chapter. Chapter four describes the fundamental breeding unit in our population, the monogamous *pair bond*. Mate fidelity and ages at first breeding are treated here. In chapter five we introduce the *helpers* with data on family compositions, helper sex ratios, and helper activities. Various analyses of *territoriality* are presented in chapter six, including the effects of family size and helper constituency upon territory size. *Dispersal* methods and patterns are discussed in chapter seven, along with our calculations of effective population sizes and levels of inbreeding in the population. Ten years' data on *reproductive success* are analyzed in chapter eight, where the important effects of helpers upon reproductive success and offspring survival are presented and discussed. Chapter nine summarizes our data on *survivorship* at all stages of Florida Scrub Jay life, thereby allowing calculation of overall life tables within the population and a discussion of their attributes. Chapter ten addresses the evolution of the cooperative-breeding system as we see it, relates our findings to studies of other communal breeders, and summarizes the directions of further research suggested by the present study.

CHAPTER TWO

Procedures

The methods and accomplishments of our study have been affected profoundly by three factors, each of which deserves special mention before we outline our basic field procedures. The three factors are (1) location of the study tract within the confines of an ecological research station, (2) the low average height of the jays' favored vegetation, and (3) the extreme tameness of most jays in our marked population.

2.1. ARCHBOLD BIOLOGICAL STATION

The study tract consists of about 350 to 400 ha of oak scrub (scrubby flatwoods, *sensu* Abrahamson et al., in press) and related vegetation formations. Located on the property of the Archbold Biological Station, the tract lies about 10 km south of the town of Lake Placid in Highlands County, Florida. The size of our tract is governed primarily by the time we have available for field work, as acceptable jay habitat essentially surrounds the study tract and forms the dominant vegetation formation over most of the Archbold Station's 1,580-ha property. In certain directions, oak scrub extends for many kilometers beyond the Station (but outside the Station it is being destroyed rapidly by man). Thus we are able to study a representative segment of a large, continuous population of Florida Scrub Jays in a habitat where they are not persecuted or in any way disturbed by humans.

The importance of habitat protection and the first-rate facilities provided by Archbold Biological Station cannot be overemphasized. Much of the information we seek requires following the lives of individuals from birth to death, and each color-marked jay, therefore, represents a major investment of our time and energy. Archbold Biological Station encourages long-term ecological research, and it is committed to maintaining the scrub environments essentially in their native condition. We doubt that any other piece of Florida oak scrub could provide better security. Furthermore, the Station provides excellent facilities, including comfortable living quarters, library, laboratory space, weather and other environmental records, and equipment useful to our study. In these and countless other ways, the study owes its greatest debt to the existence and assistance of the Archbold Biological Station, its staff and facilities.

2.2. OPEN HABITAT

Chapter three presents a summary of the habitats characterizing the study tract. We emphasize that the vegetation covering about 80 percent of the tract averages less than 2 m tall. This is extremely fortunate, as it permits us to follow individual jays, and even groups, when they fly many hundreds of meters across the scrub. It also forces the jays to nest within easy reach from the ground; almost always nests are situated below eye level and can be located and monitored with a minimum of disturbance. Finally, the low, open habitat probably contributes to the tamability of the Florida Scrub Jays, as discussed below.

The eastern edge of our study tract includes about 100 ha of scrubby flatwoods that have not been allowed to burn for many decades. As discussed in chapter three, the scrub in this patch averages several meters high. The frustrations of searching for nests, mapping territories, and following the movements of marked jays that sometimes use this tall, dense habitat serve as a constant reminder of the ease with which we can study jays living in the low, open, fire-maintained scrub of the main study tract. However, information on jays living in the unburned scrub has provided many important data regarding habitat quality, discussed in chapter three.

2.3. TAME JAYS

Although not innately tame, Florida Scrub Jays are eminently tamable. As any visitor to Archbold Biological Station is aware, many of our marked jays are spectacularly tame. Nearly all jays in our tract allow us to approach within a few meters, and about half the jays regularly perch on our hands and heads without concern. In some cases we must lift a brooding female off her nest in order to check its contents. Most jays in the study tract fly *toward* us whenever we enter the territory, although they quickly lose interest and ignore us if we refrain from interfering with their activities.

The tameness of our study animals allows us to accomplish numerous field tasks with extreme accuracy. On our monthly censuses of jays in the tract, we count *every* jay. We are confident that we discover every breeding attempt by every breeding pair in the tract. Initially timid immigrant jays quickly become tame simply through observational learning, which permits us to capture and color-ring them soon after their appearance. Our searches for dispersed birds beyond the study tract are highly successful, because the already-tamed dispersers are easily attracted to us. Finally, and perhaps most important, many kinds of behavioral observations and studies, completed or in progress, have been enhanced far beyond normal expectation by the tameness of the color-ringed jays.

We emphasize that these jays are *tamable*, not simply tame by nature. The most important condition is that they not be harassed by humans. Where pestered, or even where little exposed to humans, they are wary of human approach and scold us as they would a nest predator. As omnivorous corvids, however, they are intelligent and bold in their search for food. Where humans become food sources, Florida Scrub Jays quickly adapt to close approach. Because they dwell in tight social groups, they learn quickly from each other's actions, especially if food is a reward. It may be that the Scrub Jay's propensity for perching on the backs of wild mammals while foraging (Baber and Morris, 1980; Dixon, 1944) accounts for their frequently acquiring the habit of landing on humans when rewarded with food.

We suspect that the openness of their habitat is a primary reason for the Florida Scrub Jay's greater propensity for tameness compared with most other corvids. Individuals occupying closed habitats probably must exercise caution with every move and can never be confident that no sources of danger are near. Such a way of life must select for relative wariness compared to that of individuals that spend most of the day in full view of most of their home range. For these latter species, including the Florida Scrub Jay, open habitat provides an opportunity for individuals to watch other animals, including humans, from initially safe distances. When these animals never cause harm, and frequently produce food, wariness gives way to behavior that humans tend to view as "tameness."

Florida Scrub Jays at Archbold Biological Station were offered food, mostly peanuts and cracked corn, for several decades before our study began. As a result, many jays in the study area were tame at the onset of the project. The trait has spread through our efforts, although much variation remains within the marked population. In general, offspring of wary adults grow up wary, and offspring of tame adults quickly become tame. The jays' boldness to humans varies slightly with the season. They approach us most readily in late winter and early spring when food seems scarce. They frequently ignore us in autumn, when acorns are abundant.

We remain confident that the tiny amounts of artificial food (peanut bits) we supply the jays during periodic visits to their territories have no effect on their survival or any aspect of their social or foraging behavior. Rather, it affords us a rare opportunity to observe at close range the natural behavior and ecology of a complex bird; hence our practice of occasional feeding is well worth the price of peanuts.

2.4. FIELD PROCEDURES

Table 2.1 summarizes the ongoing field procedures, and their approximate schedules, that provide the basic life history data reported in this book. Countless supplementary projects have been undertaken, some lasting years, others only a few minutes, and these will be referred to in the text whenever appropriate. All aspects of the study, however, rely upon the raw data on genealogies, reproduction, and movement patterns generated by the basic procedures outlined in table 2.1. Remarks are necessary regarding a few of these procedures, most of which are further elaborated in other publications or later in this book.

Color-Ringing

With minor modifications in recording, we have continued with the color-ringing scheme described previously

TABLE 2.1. Basic Field Procedures, 1970–1980

Procedure	When Accomplished
Census entire marked population:	Monthly, about mid-month
Map territorial boundaries:	Annually, breeding season
Locate all nest attempts of all families:	Annually, March through June
Check contents of each active nest; estimate nest failure dates for prematurely empty nests:	Every 2–5 days after discovery
Measure, weigh, band all nestlings:	Day 11 of nestling's life
Place full color combinations on fledglings:	2–3 months after fledging
Color-ring all immigrant jays:	As soon as possible after "residency"
Color-ring, census peripheral families outside main study tract:	Irregularly, as needed
Census outlying scrub areas for long-distance dispersers:	Irregularly; thorough job in 1979

(Woolfenden, 1975), in which each jay carries two color rings and the uniquely numbered, aluminum U.S. Fish and Wildlife Service band. Ten colors are used: white (W), red (R), green (G), yellow (Y), blue (B), purple (P), orange (O), lime (L), flesh (F), and azure (A). The metal band is designated silver (S). Up to three rings fit on one leg, which, with 10 colors, makes possible 1,200 unique combinations. Each jay is named by reading the rings in sequence from top to bottom, right leg first, left leg last. A dash separates the rings carried on the right leg from those on the left leg, and *always* is included in the name. Thus, the individual bearing the band combination, "G–WS" is verbalized as "green-dash-white-silver."

To conserve unique band combinations, young jays are outfitted with a "complete set" only after they survive a few months as fledglings, although they receive the numbered aluminum band and one individual color while still in the nest. As of December 1979, 725 individuals had been uniquely color-ringed.

Sex Determination

The sex of individual Florida Scrub Jays cannot be determined reliably from plumage or external measurements. The only safe morphological feature yet found is the brood patch. This is limited to actively breeding females during a few months of the year and can be discovered only by examination of an individual in the hand. We use a suite of behavioral traits to identify females; males can be sexed positively only by their courtship displays, especially courtship feeding. The most common female-specific behavior is the "hiccup" vocalization, a mechanical, clicking sound often uttered with a stereotyped, bill-up posture by females during territorial display sequences. This call is given by helpers as well as breeding females. Most other sex-specific behavior patterns are limited to breeding pairs performing courtship, copulation (rarely observed), or nesting duties. Dominance interactions occasionally reveal the sex of a helper, particularly in cases where a helper unequivocally dominates the breeding female, a practice restricted to males.

In general, if a helper has not been heard giving the hiccup call by age 14–18 months, the helper can be assumed to be male. Unfortunately, some jays have disappeared after helping, but before we had any positive indication of sex. These jays are excluded from a few analyses, and included where appropriate in others, as noted in tables and text. In recent years we have improved at establishing the sex of helpers before their departure.

Monthly Census

Soon after the extremely sedentary nature of the jays was realized, a regular schedule of censusing was established. We have counted virtually every jay in our tract every month since April 1971. Censuses take two or three days to complete and normally are scheduled near the middle of the month.

Because the jays are tame and because we usually know the centers of activity within their large territories, our censuses are simple and direct. They consist of systematic walks through all territories, checking for presence or absence of the expected jays in each. Usually if we do not find an individual within a few *minutes* after locating its family, we are unlikely to see it ever again in the tract, although we check repeatedly for it during the census period. If it was a breeder or a bird less than 12 months old, such an absence virtually always means it is dead. If it is an older helper, however, it may have dispersed. Occasionally older helpers return after one- to three-month absences from the tract. Their return probably follows unsuccessful dispersal attempts, a topic discussed in detail in chapter seven.

Nest Checks

Critical to all facets of the study is the monitoring of every breeding attempt within every family we follow. We believe we are essentially 100 percent successful at discovering all nest attempts. This is supported both by the behavior of the breeding pairs we watch and by the fact that during thousands of hours spent afield in the scrub, we have *never* found a nest suspected to have failed before we had found it.

During the 1970s we found nearly 400 Florida Scrub Jay nests in the study tract. Of these, 24 percent were found before the first egg was laid, 48 percent before incubation began, 70 percent before day 2 of the 18-day incubation period, and 99 percent before hatching. Only 2 nests were not found until shortly after hatching, and 1 aberrant nesting was not discovered until a day or two after fledging (case history 30). In recent years no nest has gone undiscovered past early incubation.

Nest finding basically involves noting a morning absence of the breeding female, then following her once she is located off the nest. Nests found before egg-laying are located by following breeders that are carrying sticks or nest lining. (Occasionally we can present such items to the jays, who take them from us and fly to the nest.) Male breeders feed their incubating mates on the nest, and these feeding flights also help us locate nests.

We check nests every few days throughout their cycle to document survival rates of eggs and nestlings. To minimize the risk of nest predation caused by our visits, we keep them brief and perform them near midday when predators are probably least active. We attempt to leave no trace of our paths to the nests. We have not yet performed analyses to determine whether our visits affect nest survival, but we have no suspicions that they do.

The only major disturbance we create during the nest cycle occurs on or near day 11 after hatching, when nestlings are banded, measured, and weighed. During these visits we remove all young from the nest at one time, and keep the young quiet and obscured. Even breeding jays that initially tend to scold at our approach become quiet during most of the banding process.

Territory Maps and Density Calculations

Each breeding season, usually in April or May, we map the territories of all families in the main study tract. Reference copies of recent aerial photographs are carried in the field and are used in plotting boundaries indicated by territorial disputes between neighboring families. We are

19

able to determine many boundary segments by observing naturally occurring disputes. Where necessary we instigate disputes by attracting neighboring families to the likely region using food, our own imitations of their calls, or a tape recording of territorial vocalizations.

Because of several features in the biology of Florida Scrub Jays, our territorial maps are extremely accurate. The families are sedentary, and boundaries typically shift only slightly from year to year. More important, boundaries themselves are only several meters wide where they cross a patch of oak. Ponds or low palmetto depressions, visible on the aerial photo, frequently form large portions of the boundaries. The jays' territorial display often involves long, undulating flights along the boundary, thereby showing us exactly where to draw the boundary line on the map. When neighboring families meet at close range on the ground, mutual displaying between jays of like sex can occur. In only a narrow zone, exactly at the territorial boundary, is neither bird clearly dominant over the other.

Each territory map reproduced in Appendixes B through K represents a painstaking, independent, and accurate plotting of each such boundary segment. The field maps are traced onto a master copy of the aerial photograph, on which the ponds and forested habitat patches are outlined in standardized fashion. Where these unused patches form boundaries, we trace the entire patch (with standardized shape) as a boundary. In calculating territory sizes by means of a compensating polar planimeter we subtract *all* area contained within these unused patches, including such patches entirely within a territory as well as those along boundaries. Thus, our territory size measures cannot be used directly in calculating overall *density* across the entire habitat. They are, however, directly comparable between years, because the same total area is removed as unused habitat regardless of the exact position of the boundaries.

Population densities are calculated four times annually, in March (onset of breeding), June (peak of fledgling abundance), September (peak of dispersal by helpers), and December. Densities reported in this book refer to the core area of the tract, consisting of about 225 ha of favored habitat. The exact area delimited in our calculations varies each year according to territory boundaries because we measure density in whole territory units. Thus, we calculate the total area enclosed by the perimeters of a pre-designated set of core territories. Within this set of territories we count (1) number of breeding pairs, which almost always equals the number of territories, (2) number of breeding individuals, defined as experienced breeders or novices that have formed a clear pair bond, (3) number of helpers older than age 1 year, and (4) number of juvenile helpers less than 1 year of age. Together, these categories comprise all jays in the known core area of the tract, providing density measures for total numbers or any component thereof. Density is calculated in birds per 40 ha (100 acres).

Peripheral Censuses

Through the years, a number of jays who were hatched, banded, and raised in the study tract have dispersed successfully to become breeders outside of it. Undoubtedly, a few of these have escaped our detection, especially any that disperse extremely long distances. However, we feel we have discovered a high proportion of these individuals, either through periodic censuses of the surrounding scrub or via information given to us by other observers working elsewhere on the Archbold Biological Station property or adjacent areas. The tameness of our jays is of great advantage in this regard, because most dispersed jays (still car-

21

rying their color-rings) are attracted to any human entering their territory. We are informed of such encounters, even by local residents away from the Station. Any such individual is identified positively by us, and the distance away from its natal territory is calculated in average territory units.

During the 1979 breeding season we made a careful search for banded jays in a wide area surrounding the tract. We studied aerial photographs to locate patches of appropriate habitat, and we traversed them by truck, playing tape recordings of Scrub Jay territorial calls to attract the resident groups. In the process we censused every jay family on the biological station property (ca. 1,000 ha of scrub) and an additional 400 ha of adjacent jay habitat. We encountered over 200 families in the contiguous scrub outside our tract. Average family size exactly matched that within our tract, and jay density, corrected to exclude large patches of inappropriate habitat, also matched that in our tract. As discussed in chapter seven, only nine dispersers were located, and all but one of these had been known to us before our intensive search.

2.5. TABULAR DATA

Because many of our data can be used in ways other than as used in this book, we include tables presenting the raw data for numerous analyses discussed in the text. In general, the tables include summary rows or columns to facilitate rapid perusal of the main results.

It is important to note that in many instances several data tables incorporating similar or identical measures show slight differences between them in the raw numbers. This arises because of the conflict between (1) attempting to use maximum sample sizes and (2) eliminating equivocal data elements or those whose inclusion would produce some form of sampling bias. Some results (e.g., the fates of particular fledglings, ages of certain helpers, survival of certain peripheral nests, and small sample of nests studied in 1969, etc.) are appropriately included in certain analyses but excluded from others. We have tried to clarify each sampling procedure and data base through footnotes or text explanations. In all cases, the equivocal data are few, and we consider any such discrepancies to reflect our careful attempts at avoiding biases in sampling procedure.

CHAPTER THREE

The Scrub Jay in Florida

SYNOPSIS

The Florida Scrub Jay is an isolated, relict population of a species with a wide geographic range in western North America. It occurs only in the botanically distinct Florida oak scrub, a rare, scattered habitat whose islandlike distribution is being reduced further by man. The jays' use of microhabitats shows obligatory reliance on oaks, especially those growing in short, open scrub maintained by periodic fire. Little suboptimal habitat exists—where the scrub becomes taller or dominated by pine, Blue Jays replace Scrub Jays. Reproductive success in any but the short, open scrub is poor. All available habitat is constantly occupied, and population densities of breeding Florida Scrub Jays are extraordinarily stable from year to year. Various measures show the population to be at carrying capacity, densely crowded with breeding birds. Mature, nonbreeding individuals are present in many territories, so that every new breeding space is filled instantly. Most of this monograph is a description of the interplay between these breeding and nonbreeding birds. Cooperative behavior has evolved in response to this interplay, ultimately because of habitat saturation. Southwestern North America includes a wide variety of habitats acceptable to Scrub Jays. There Scrub Jays are not cooperative breeders.

3.1. GEOGRAPHIC HISTORY

The Florida Scrub Jay is a geographically isolated, morphologically distinct population of a species whose distribution in western North America extends from the Oregon chaparral to the pine-oak woodlands of southern Mexico (see Pitelka, 1951; figure 3.1). The Scrub Jay is the northernmost of three species in its genus, and the only one whose range extends significantly beyond the mountains of Mexico. The Mexican, or Gray-breasted, Jay (*Aphelocoma ultramarina*), a species whose complex behavior is being studied extensively by J. L. Brown (1963, 1970, 1972, 1974), is largely restricted to lowland and mid-montane evergreen forests of the Mexican plateau. The little-known Unicolored Jay (*A. unicolor*) is a rare endemic in the montane oak-pine and cloud forests from southern Mexico to Honduras. Its social system is unknown.

The Florida race of the Scrub Jay is separated by about 1,600 km from its nearest relative (*A. coerulescens texana*), which occupies the semiarid oak woodlands of central Texas. In view of numerous morphological, behavioral, and social differences between the Florida form and its western relatives, a strong case can be made for elevating the Florida population to full species status. For present purposes, however, we leave open this taxonomic point.

As postulated by Pitelka (1951), Neill (1957), and others, the relict Florida population probably originated with an easterly expansion of a western xeric biota sometime in the late Tertiary or early Pleistocene. Changing climatic patterns of the Pleistocene left numerous modern, peninsular Florida endemics restricted to the limited, patchy remnants of this ancient habitat. Neill (1957) provides a list of these species. As noted by Neill and by numerous other naturalists (e.g., Allen, 1871; Bendire, 1895; Bent, 1946; Sutton, 1949), the Florida Scrub Jay is one of the most typical elements of this distinctive community; the bird is virtually unknown from any other Florida habitat (Howell, 1932).

The oak scrub of Florida is confined to local areas, pri-

25

FIGURE 3.1. Total range of the Scrub Jay (*Aphelocoma coerulescens*), traced from Pitelka (1951).

marily recent or ancient dunes, where excessively drained sandy soils and highly seasonal rainfall patterns produce xeric growing conditions. In these areas the vegetation is dominated by a few species of stunted, woody perennials, chiefly sclerophyllous oaks (*Quercus* spp.) of several species, as described below. The height of most plant life does not exceed two meters, and in most places a human walking through the scrub has an unobstructed view of several kilometers over the tops of oaks and palmetto thickets. At present, as shown in Jackson (1973), this relict, desertlike habitat occurs only as isolated small patches in the central peninsula of Florida. Westcott (1970) produced the first detailed map of the Florida Scrub Jay's total native distribution. His map is reproduced here as figure 3.2, modified with additional localities after Cox (1983). The islandlike nature of this distribution reflects the jays' confinement to a rare and local vegetation formation. Unfortunately, the oak scrub grows on well-drained soil that is also an ideal substrate for citrus groves and housing developments. As a result, the Florida Scrub Jay's overall range continues to decline at an alarming rate as man destroys the final vestiges of its endangered habitat (Kale, 1978; Cox, 1983).

This peculiar geographic history of ancestral range expansion followed by habitat fragmentation and exceptional habitat specificity (see below) suggests that the Florida Scrub Jay population has contracted drastically from a relatively recent, more widespread Pleistocene range. Presumably, local populations were subjected to increasingly crowded surroundings as their habitat disappeared. This historical picture is of special importance to the Florida Scrub Jay story. Habitat saturation and associated intraspecific competition for breeding sites are commonly postulated as key precursors to delayed breeding, group living, and helping behavior in birds (e.g., Brown, 1969, 1974, 1978a; Emlen, 1978, 1982a,b; Gaston, 1978a; Selander, 1964; Woolfenden, 1976a). Data we have obtained regarding patterns of habitat use and population densities provide field evidence that intense intraspecific crowding, particularly with regard to breeding space, does indeed characterize the Florida Scrub Jay. These observations will be presented after a brief introduction to the general features of the habitat within which our subjects live.

3.2. THE SCRUB HABITAT

From the Scrub Jay point of view, about seven distinct microhabitats occur in the core area of our study tract at the Archbold Biological Station. These different vegetation

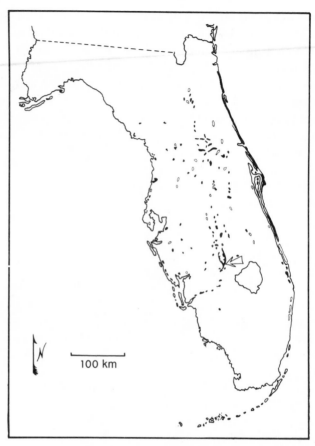

FIGURE 3.2. The total native range of the Florida Scrub Jay (*Aphelo-coma c. coerulescens*) (black areas) includes a narrow strip along the east coast of the peninsula, a few isolated areas along the west coast, and clusters of relict sand dunes in the central peninsula. Some patches of apparently suitable habitat (small unshaded areas) lack the jays. The Archbold Biological Station is situated near the south-ern tip of the jay's interior range (arrow), just northwest of Lake Okeechobee. (Adapted from Westcott, 1970, and Cox, 1983.)

zones are associated with elevational differences of only a few meters, from the tops of relict sand dune ridges down to old interdunal depressions of varying depths. As shown by the aerial photograph of the study tract (figure 3.3), these distinct zones usually meet along remarkably narrow ecotones. The habitats described here are diagramed in figure 3.4 and mapped in Appendix A. More extensive botanical descriptions of the overall formation may be found in Laessle (1942) and Abrahamson et al. (in press).

1. Oak scrub occupies the dune tops (figures 3.5, 3.6). This stunted vegetation consists mostly of .5- to 1.5-meter-high scrub oak (*Quercus inopina*), with scattered, open, taller copses of sand live oak (*Q. geminata*) and solitary Chapman oaks (*Q. chapmanii*) interspersed throughout. These oaks are evergreen, with a simultaneous leaf-drop confined to a few weeks in April, followed by immediate flowering and growth of new leaves. Thus the sclerophyllous oak scrub is an evergreen habitat except for a few leafless weeks in the heart of the Scrub Jay nesting season (see chapter eight). A few stands of sand pine (*Pinus clausa*) and rosemary (*Ceratiola ericoides*) occur on the higher ridges, often mixing with the oaks. Even the densest oak patches contain numerous openings that show the underlying bare, white sands (see Abrahamson et al., in press, for soil formations), only sparsely dotted with herbaceous annuals. Bird densities are very low in open oak scrub, with only about 50 pairs per 40 ha during breeding (Woolfenden, 1969). Besides Scrub Jays, the typical birds of oak habitat are Northern Bobwhite (*Colinus virginianus*), Loggerhead Shrike (*Lanius ludovicianus*), Rufous-sided Towhee (*Pipilo erythrophthalmus*), and in winter, Palm Warbler (*Dendroica palmarum*). Towhees, which constitute about half of all breeding pairs, are the commonest scrub birds throughout the year.

2. On the well-drained sands of gradual interdunal slopes

29

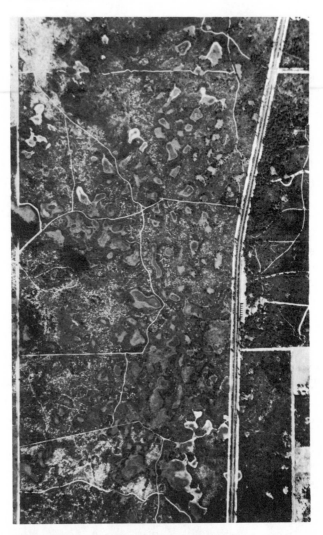

FIGURE 3.3. Aerial photograph of the jay study tract and some surrounding areas at the Archbold Biological Station. The numerous, small, irregular pale zones are grassy depressions, dry when photographed, but sometimes flooded. Sand roads and fire lanes appear as white lines crisscrossing the study tract. The main laboratory building of the Archbold Biological Station appears near the center right edge of the photograph, just right (east) of a railroad track. Gross habitat features of the study tract are mapped in Appendix A. Photo courtesy of Archbold Biological Station.

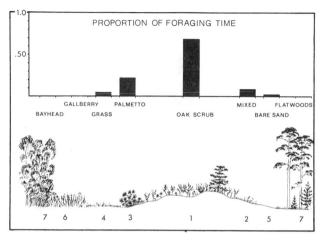

FIGURE 3.4. Simplified habitat profile of the jay study tract (see text for detailed descriptions). Graphed above are Florida Scrub Jay foraging times in each of these habitats.

FIGURE 3.5. Prime oak scrub for the jays in the study tract showing average height of the oaks. In foreground, saw palmetto and wiregrass; in left background, two sand pines. (Habitat 1 in text and figure 3.4.)

FIGURE 3.6. Prime oak scrub showing typical bare sand patch. Left, scrub palmetto at base of rosemary. (Habitat 1 in text and figure 3.4.)

and in some lower, flat areas (figure 3.7) grows a mix of low saw palmetto (*Serenoa repens*); scrub palmetto (*Sabal etonia*); sparse and stunted oaks, including dwarf live oak (*Quercus minima*); and various grasses, especially wiregrass (*Aristida stricta*). Prickly-pear cactus (*Opuntia compressa*) thrives here, as well as on the oak-dominated dune tops. These stretches are very open and easily walked, with much bare sand and few plants exceeding .5 m in height. The birds most typically associated with these zones are Eastern Meadowlark (*Sturnella magna*) and Bachman's Sparrow (*Aimophila aestivalis*).

3. Tall and densely interwoven saw palmettos form tight, narrow belts around the edges of low depressions, particularly those grassy zones that are seasonally inundated (figure 3.8). These palmetto thickets, which sometimes reach 2 m in height, are virtually impenetrable to humans and are used by several bird species for night roosting (e.g., Red-winged Blackbird, *Agelaius phoeniceus*).

FIGURE 3.7. Gradual interdunal slope, showing expanse of low-growing saw palmetto and wiregrass, which is used infrequently by the jays. Background forest includes the thick bayhead (right) and a little used, dense stand of pines (left). (Habitat 2 in text and figure 3.4.)

4. Dense stands of grasses and sedges (chiefly *Panicum abscissum* and *Andropogon* spp.) and a few woody perennials (e.g., *Hypericum edisonianum*) occupy the lowest points in most interdunal depressions (figure 3.8). Usually these well-defined grassy patches are dry or only slightly damp. During late summer and fall, however (August through November, after the summer rains), they may hold a meter or more of standing water. The Common Yellowthroat (*Geothlypis trichas*) occupies the margins of these seasonal ponds, and when full of water the ponds attract a few Sandhill Cranes (*Grus canadensis*), Mottled Ducks (*Anas fulvigula*), and herons.

5. Patches of bare sand occur throughout the study tract (figure 3.9). These are scattered among the higher ridges, along vehicle trails and fire lanes, and along the edges of certain grassy depressions. Some interdunal low spots are rimmed with extensive borders of bare white sand, on which

FIGURE 3.8. Grassy interdunal depression ringed with tall, dense saw palmetto. These large depressions often form segments of territorial boundaries (see Appendixes B–K). (Habitats 4 and 3, respectively, in text and figure 3.4.)

grow a few wisps of hardy grasses. Sandy openings are a principal nesting microhabitat for the Common Ground-Dove (*Columbina passerina*) and Common Nighthawk (*Chordeiles minor*).

6. Thickets of gallberry (*Ilex glabra*), fetterbush (*Lyonia lucida*), wax myrtle (*Myrica cerifera*), and tall grasses occupy some of the moist low areas (figure 3.10). Slash pines (*Pinus elliottii*) become numerous around these and other low areas; otherwise these pines are thinly dispersed over the study tract. Northern Mockingbirds (*Mimus polyglottos*) often reside in thickets; Mourning Doves (*Zenaida macroura*) nest in the pines.

7. Two principal kinds of taller forest are present as small patches within the study tract. "Bayhead," a lush, subtropical-looking forest 8–12 m high, grows in two low depressions in the tract, each only a few hectares in area

FIGURE 3.9. Jay foraging on bare sand patch typical of substrate beneath oak scrub. Scrub oaks (*Quercus inopina*) surround the opening. (Habitat 5 in text and figure 3.4.)

FIGURE 3.10. Gallberry thicket at edge of tall-grass depression. Slash pines (background) often are numerous around these low areas. (Habitat 6 in text and figure 3.4.)

(figure 3.11). This dense, evergreen formation consists mainly of several species of bay (*Persea borbonia, Gordonia lasianthus, Magnolia virginiana*). "Low flatwoods" are thick stands of old, 15- to 20-meter-high slash pines growing over a grassy understory interspersed with saw palmetto thickets (figure 3.12). These pine forests loosely border the study tract at several points, especially along the eastern edge where fire prevention has allowed the pines to form an almost closed canopy (see Appendix A). Only in these true forest microhabitats does bird diversity exceed a few species. The characteristic birds of the forests include Eastern Screech-Owl (*Otus asio*), Chuck-will's-widow (*Caprimulgus carolinensis*), Red-bellied Woodpecker (*Melanerpes carolinus*), Great Crested Flycatcher (*Myiarchus crinitus*), Carolina Wren (*Thryothorus ludovicianus*), White-eyed Vireo (*Vireo griseus*), Northern Cardinal (*Cardinalis cardinalis*), and most important to this study, the Blue Jay (*Cyanocitta cristata*). Blue Jays do not use the open, scrub areas; as shown below, Scrub Jays avoid the forests.

Three hawks (Red-tailed Hawk, *Buteo jamaicensis;* American Kestrel, *Falco sparverius;* and recently, Cooper's Hawk, *Accipiter cooperii*) and another owl (Great Horned Owl, *Bubo virginianus*) reside on the Station. Other hawks occur as migrants or winter residents (e.g., Sharp-shinned Hawk, *Accipiter striatus;* Northern Harrier, *Circus cyaneus;* Merlin, *Falco columbarius*).

3.3. FLORIDA SCRUB JAY HABITAT USE

Since the beginning of the study, our impression has been that the Scrub Jays are predominantly associated with the oak scrub (habitat 1, above), despite the presence in

FIGURE 3.11. Lush bayhead forest (8–12 m tall) in low, poorly drained area along west boundary fence in study tract (see Appendix A). Jays forage in bordering oaks (foreground), but do not penetrate the forest. (Habitat 7 in text and figure 3.4.)

FIGURE 3.12. Slash pine flatwoods with open palmetto understory and virtually no oaks. Note trunks blackened by recent fire. (Habitat 7 in text and figure 3.4.)

the tract of many other identifiable vegetation types. From October 1976 through September 1977, Jerre A. Stallcup gathered data on the use of all these habitats in order to test this impression. She followed marked jays of several different families during a composite day each month for 12 consecutive months. Each month's composite day consists of all the clock hours of a typical day drawn from two to six calendar days (mean = 2.5 days). All the habitats listed above were available for use by the jays under observation, although not necessarily in the percentage typical of the tract as a whole. Stallcup's data are summarized in figure 3.13 and in the graph included in figure 3.4.

Stallcup recorded only the foraging sites of jays under observation. However, with rare exceptions, the jays conduct *all* their nonforaging activities in the same habitats where they forage. A study of their daily and annual time budgets by DeGange (1976) shows that jays spend about 40 percent of the daylight hours foraging and about 50 percent perching. Nonforaging jays nearly always perch in the same vicinity and habitat where their family is, or recently has been, foraging. We therefore treat the foraging data in figure 3.13 as accurate reflections of the overall, daytime habitat use pattern in our Florida Scrub Jay population.

Figure 3.4 shows that two-thirds of a jay's typical day is spent in the oak scrub. The percentage of time spent in oaks never drops below about 50 percent at any time of the year (figure 3.13), and reaches a peak of about 90 percent during late summer, when acorns are ripening. Furthermore, all remaining habitats used by the jays are uniquely associated with the open scrub formation as portrayed in figure 3.4. The jays avoid forested habitats entirely, even the sparsely wooded low flatwoods that form a significant

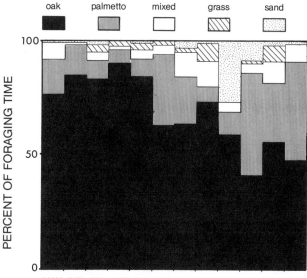

FIGURE 3.13. Monthly percentages of Florida Scrub Jay foraging time spent in each of five scrub microhabitats. Oak scrub clearly is favored foraging zone in all seasons.

proportion of the property of Archbold Biological Station and that constitute the dominant vegetation of central Florida. The thick, rank stands of gallberry and fetterbush, also common in low areas of the study tract, are spurned as well. Beyond Archbold Biological Station, Florida Scrub Jays are entirely absent from prairies and marshes that make up virtually all of central Florida's nonforest habitat. In brief, the Florida Scrub Jay relies heavily upon the oak scrub association, a reliance that appears to be obligatory despite the abundance of other, seemingly usable Florida habitats.

We have evidence that the jays even reject some patches of oak scrub that appear marginally usable. In the early 1940s, Amadon (1944) found Florida Scrub Jays living throughout oak scrub habitats east of the main building on the original property of the Archbold Biological Station. Prior to that time, this area had experienced periodic natural burning, but was undisturbed by man. Early photographs (ca. 1930–1935, Archbold Biological Station archives) show that it was dominated by low, open oak scrub. No extensive fires have occurred since 1941, when the Archbold Station was founded. Successional changes over the 35 subsequent years have resulted in denser, taller scrub with thick leaf litter, abundant ground-story vegetation, and more mature slash pines (figure 3.14; see description in Woolfenden, 1969, under "Scrubby Flatwoods"). Between 1969 and 1980, the Scrub Jay population in this patch of scrub dropped from four pairs (1969, 1970) to one or zero (1977–1980; figure 3.15). In 1979 we even watched the one remaining family literally fight its way westward to gain a small piece of more open scrub in which they eventually nested. The jays seem to recognize the taller scrub as suboptimal. Indeed, as shown in table 3.1, fledgling production and fledgling survival have been significantly lower in this tall scrub than in the open, periodically burned scrub of the main study tract (see habitat map in Appendix A). Reduced productivity in the tall scrub is accompanied by a greater-than-usual number of nest attempts, which in turn results in more eggs laid per pair per season (table 3.1). Both of these indicate an increase in nest predation rate compared to the open scrub. We suspect that the presence of Blue Jays in the taller scrub may account for much of the decrease in success by Scrub Jays in this habitat (see 3.7. "Conclusions").

40

FIGURE 3.14. Tall, dense vegetation characteristic of unburned oak scrub found east of railroad at the Archbold Station. Jay populations in the tall, dense scrub declined to near zero during the 1970s (see figure 3.15).

3.4. FOOD

Typical of the family Corvidae, Florida Scrub Jays are omnivores. Insects, principally orthopterans and lepidopteran larvae, appear to form the bulk of the animal diet over most of the year. These and other large arthropods (including spiders, scorpions, centipedes, and millipedes) are picked or probed from the low vegetation or the ground. Fast-flying or fossorial prey are rarely taken. The jays most frequently hunt for food by hopping along the bare sand under scrub oaks, or by jumping from shrub to shrub within

41

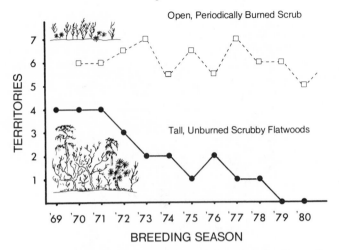

FIGURE 3.15. Extirpation of a viable Scrub Jay population from tall, unburned scrub shown in figure 3.14. See also Appendixes B–K, northeast of laboratory building. As a control, dashed line shows territorial density in a similar area on the main study tract, directly west of the unburned, tall scrub (west of railroad tracks in Appendixes B–K).

the oak foliage or palmetto fronds, carefully examining leaves and darting after any startled creature attempting to escape.

A variety of small vertebrates are eaten whenever encountered. Vertebrate prey we have witnessed being taken include frogs and toads (*Hyla femoralis, H. squirella,* rarely *Bufo quercicus,* and unidentified tadpoles), lizards (*Anolis carolinensis, Cnemidophorus sexlineatus, Sceloporus woodi, Eumeces inexpectatus, Neoseps reynoldsi, Ophisaurus compressus, O. ventralis*), small snakes (*Thamnophis sauritus, Opheodrys aestivus, Diadophis punctatus*), small rodents (*Sigmodon hispidus, Peromyscus polionotus, Rattus rattus* young), downy chicks of the Bobwhite (*Colinus virginianus*), and a fledgling Common Yellowthroat (*Geothlypis trichas*). Florida Scrub Jays occasionally rob other birds of their eggs or young, but our

42

Table 3.1. Reproductive Success and Survival of Young Jays in Two Kinds of Scrub Habitat, 1969–1978

	Scrub Quality[a]	
	Dense, tall	Open, low
Pairs (N)	40	193
Eggs	221	960
Fledglings	54	402
Yearlings	12	147
Eggs per pair	5.53	4.97
Fledglings per pair	1.35	2.08
Yearlings per pair	0.30	0.76
Survival of nest contents[b]	0.24	0.42
Survival of fledglings[b]	0.22	0.37

[a] See text for full descriptions of dense, tall (unburned) versus open, low (periodically burned) oak scrub habitats.
[b] Survival of nest contents measured as proportion of eggs surviving to produce fledglings; survival of fledglings measured as proportion of fledglings surviving to age 1 year.

observations indicate this is a rare event, which contrasts sharply with the behavior of Blue Jays, who appear to be habitual nest robbers. A few times we have led Scrub Jays to bird eggs to study their reactions. Often they go to the eggs immediately and seize them without breaking them. Only rarely do they break open the shells and eat the contents. Instead they fly off and hide the eggs in the vegetation or sand. We do not understand these actions, but breeding-bird densities in the oak scrub are so low that the jays cannot rely on bird eggs or young as a regular food source. Intraspecific nest robbing occurs, but we suspect this also is a rare event. Certainly their strict territoriality would prevent most interfamilial nest robbing. In 1980, a replacement female, who was pairing with a recently widowed male, was seen feeding on a dead nestling age 7 days, which the male had removed from his nest after it had died from overnight exposure.

Frequently we have seen Florida Scrub Jays outside of the study tract pick at roadside carrion. Once a jay in the study tract was observed feeding on the remains of a raptor-killed rabbit.

Acorns form the principal plant food. The jays spend a considerable proportion of their day gathering ripening acorns from all the scrub oak species during late summer and fall (DeGange, 1976). Many of these they consume immediately, but the great majority are cached in the sand, husks intact. These are recovered, husked, and eaten throughout the rest of the year. We remain uncertain as to how important these cached nuts are to the annual nutrition of the jays, but in all seasons we regularly see jays throwing sand as if searching for cached acorns. A few such searches produce a reward, although most do not. Every fall we observe the jays burying acorns actively until none remains on the shrubs or visible on the ground. The jays consume tremendous numbers of acorns during the autumn, and they regularly find buried ones during the other seasons. Hence, we suspect that acorns do form a necessary and year-round vegetable staple.

The jays occasionally pluck and eat other small nuts, fruits, and seeds when available (e.g., hickory nuts, *Carya;* palmettos, *Serenoa;* tread-softly, *Cnidosculus;* briars, *Smilax;* blueberries, *Vaccinium;* gallberry, *Ilex;* rosemary seeds, *Ceratiola*). However, the berries seem to be more experimental "desserts" than regular items in the diet. Weed and grass seeds are rarely if ever eaten. Corn, peanuts, sunflower seeds, and many other human-offered foods are readily consumed once the jays have learned about them.

44

3.5. POPULATION DENSITY

Table 3.2 summarizes Florida Scrub Jay population densities from 1971 to 1979 within about 225 ha of prime scrub habitat. Several aspects of these data bear directly on the question of habitat saturation and strongly suggest that the jays in our population currently exist at, or very near, their maximum potential breeding density.

1. The density of breeding Florida Scrub Jays varied over nine years from 3.59 to 4.04 pairs per 40 ha. For one square mile (259 ha) of scrub these figures translate to a standard deviation of about one pair (1.04 prs.) around a mean of 24.6 pairs. The coefficient of variation in breeding pair density (4%) is exceedingly low compared to all other passerines whose populations have been monitored (the best studies summarized by Ricklefs, 1973).

2. As shown in figure 3.16, the breeding density of jays in our tract shows no relation to total population density. The absence of such a relation is striking, because overall jay density measured at the onset of breeding has varied considerably, from 11.08 to 14.55 birds per 40 ha (see table 3.2; CV = 10%). Because the Florida Scrub Jay exhibits delayed breeding (see chapter four), we also tested for a correlation between breeder density and the density of jays one and two years previous to each breeding season (figure 3.16). Again, no relationship emerges. It appears that in our population the density of breeding pairs remains essentially fixed. The extra birds must await, and compete for, their entry into a saturated breeding population.

3. We find a significant negative relation between territory size and mean breeder density (figure 3.17), even though both measures vary only slightly from year to year. This relationship exists because the rare establishment of

45

TABLE 3.2. Florida Scrub Jay Population Densities, 1971–1979

	Breeding Season									Mean ± S.D.
	'71	'72	'73	'74	'75	'76	'77	'78	'79	
Pairs/40 ha	3.73	4.04	3.88	3.59[a]	3.79	3.72	3.61	3.83	4.00	3.80 ± 0.16
Jays/40 ha	13.44	11.94	11.46	11.08	11.18	14.55	12.12	12.43	13.61	12.42 ± 1.20

NOTE: Data from about 225 ha of scrub in core of study tract, censused in March.
[a] 1974 includes one bigamous trio.

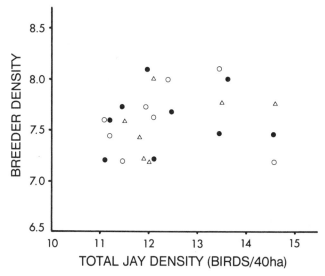

FIGURE 3.16. Breeder density plotted against total jay density in the study tract during the same years (closed circles). Also plotted are breeder densities versus total densities in preceding one (open circles) and two (open triangles) years. All densities measured in March at the onset of breeding. No correlations are evident.

a wholly new territory (see chapter seven), a few of which combined can appreciably raise the breeder density, occurs only at the expense of previously existing territories. New territories are not created in previously unused habitat. In any given year, virtually no suitable Scrub Jay habitat exists outside of an occupied and defended territory. Once again, the data clearly show that the Florida Scrub Jay world is one fully saturated with breeding pairs.

To examine relative densities of breeders and nonbreeders more closely, and to illustrate the fundamental relationship between these two classes of jays, we present their seasonal fluctuations as calculated four times annually over nine years (figure 3.18). In these data, first-year birds (juveniles, or prereproductives), older nonbreeders ("helpers"

47

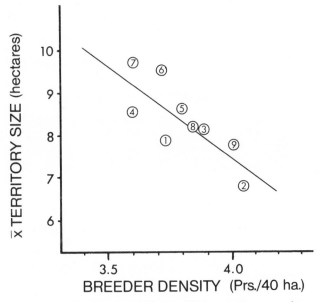

FIGURE 3.17. Negative correlation $(-.77)$ between mean territory size and breeder density, reflecting saturated habitat. Circled numbers, 1–9, represent the years 1971–1979; $r^2 = .59$.

of reproductive age), and breeders are separated. Overall jay density shows strong seasonal peaks each June as a cohort of new fledglings joins the population. Wide seasonal and annual variation exists in the density of fledglings (CV = 43%) and older nonbreeders (CV = 36%). As mentioned above, breeder density is virtually constant (CV = 4%).

Figure 3.18 depicts a densely packed population of breeders coexisting with a variable excess population of nonbreeders that are capable of breeding as soon as opportunities arise. The bulk of this monograph can be described as a detailed analysis of the dynamic interplay between these two populations.

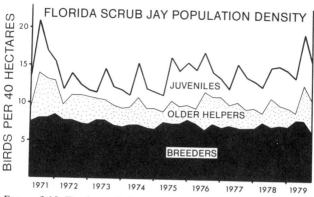

FIGURE 3.18. Total population density of Florida Scrub Jays in the study tract between March 1971 and December 1979, calculated quarterly (March, June, September, December). The component densities of three classes of jays are distinguished. Breeder density shows remarkably little variation throughout nearly a decade (see text). Annual reproduction causes a peak in juvenile density each June; at the same month, the previous year's "juveniles" become classified as "older helpers."

3.6. WESTERN SCRUB JAYS

Already emerging is the broad picture of a Florida population confined to a relict habitat, so crowded and yet so long lived that the offspring must "queue" for breeding space during several years after fledging. As we will show in later chapters, the offspring's evolution into helpers is intimately tied to this waiting game. The feature of Scrub Jay biology that makes this story doubly interesting and illustrative is that widespread populations of Scrub Jays in western North America show very different social characteristics. Various scientists have observed and reported on the breeding biology of western Scrub Jays in different parts of the range, and *not a single observation of helping by a yearling or older jay has ever been reported,* either in the literature or through conversations with us.

As best we know it, western Scrub Jays in their continental range follow breeding patterns typical of most Temperate Zone passerine birds (Pitelka, 1951; Brown, 1963; Westcott, 1969; Stewart et al., 1972; Verbeek, 1973; Atwood, 1978, 1980a,b). Monogamous pairs breed in the spring; offspring are cared for in their parents' territories until independent, then they disperse permanently from their home ground. Juvenile birds travel in loose bands during their first autumn and winter, and breeding apparently commences at age 1 year. (Detailed studies of individually marked jays are still needed to document this pattern.) The key difference between the western and Florida populations is clear. Young jays do not remain in their natal territory beyond their first summer of life in any but the Florida race. Western Scrub Jays are not cooperative breeders.

J. L. Atwood (1978, 1980a,b) recently studied the Scrub Jays of Santa Cruz Island, the largest of the Channel Islands of California. Here a distinctive subspecies (*A. coerulescens insularis*) is restricted to one 248-km^2 (96-square-mile) island about 30 km from the mainland (see Pitelka, 1951). Although some yearlings do breed in this crowded population, Atwood found a significant tendency toward delayed breeding. Nonbreeding individuals up to several years old form roving flocks that dwell in suboptimal, ridgetop habitats between the permanently defended territories of breeders. The significance of this finding will be discussed in chapter ten. The relevant point here is that even in an environmentally crowded population of western jays in which delayed breeding *has* evolved, cooperative breeding *has not*.

In contrast to Florida, western North America has a variety of habitats acceptable to Scrub Jays. In Florida, scrub often abuts the forested habitats of Blue Jays. No comparably sharp habitat replacement occurs in the West, except where Scrub Jays range up to the primarily coniferous elevations inhabited by Steller's Jay (*Cyanocitta stelleri*). Below about 2,000 m elevation, western Scrub Jays occupy many types and heights of woodland and scrub. Although they do show a preference for those containing oaks, they are not restricted to oak scrubs. Bent (1946) mentions Scrub Jays ranging from the lower Sonoran Desert up to transition zone pine-oak woodlands. Bent adds juniper hillsides, sagebrush and mountain mahogany slopes, willow ravines, orchards, and digger-pine chaparral to the wide variety of oak-dominated habitats cited as Scrub Jay habitat in various parts of their western range. All western subspecies are more arboreal, and seem to be more capable fliers, than their Florida counterparts (personal observations).

Any scenario we develop for the adaptive evolution of helping in the Florida Scrub Jay must account for the absence of this behavior in the West. This is both a blessing and a curse. The blessing is that we are granted a natural experiment of sorts. We can test for certain, key ecological differences between sister populations that might be related to the existence of cooperative breeding in one and not in the other. The curse is that we know so little about the demography of the western populations that we cannot yet use the experimental design. We trust that this frustration will be short lived, and we hope that statements and guesses we make at this point will encourage ecologists in western North America to study critically the Scrub Jays in that region.

3.7. CONCLUSIONS

We strongly suspect that the major features of the Florida Scrub Jay social system ultimately result from their extreme habitat specificity and the restricted, islandlike distribution of their favored habitat. Historical patterns of climate and vegetation in eastern North America relegated the Scrub Jay to tiny, fragmented, and sharply bounded patches of a rather homogeneous habitat. Few other North American birds are so habitat-restricted, including Scrub Jays in the West.

In subsequent chapters we show that permanent territoriality and considerable longevity also characterize our study population. The inevitable result of combining these demographic features with severe habitat limitation is intense and ongoing competition for living space. This being the most critical and limited resource necessary for breeding, we focus on this competition and its apparent results throughout this monograph.

Most bird populations that have been censused carefully over extended periods show appreciable annual fluctuations. Lack (1966) and Ricklefs (1973) review the best of these studies and show that typical coefficients of variation in population density vary from 20 to 40 percent. (Although few of these studies explicitly separate breeding from nonbreeding individuals, most estimates of variation imply or include values for breeding individuals.) No bird species ever studied shows population constancy even approaching that reported here (table 3.2; figure 3.18) for breeding Florida Scrub Jays. As detailed in later chapters, this virtual absence of annual fluctuation is a direct result of the constant presence of fully mature potential breeders "waiting their turn." These birds fill each breeding slot, or create their own, the moment an opening appears. We can conceive of no more direct evidence that a natural population is thriving at maximum potential density. We cannot overemphasize this point, because every feature of the jays' life history, and any theoretical analysis of strategic options, must be viewed in this context.

The ultimate reasons for the Florida Scrub Jays' restriction to scrub oak patches still are unclear. Some evidence exists that they require acorns in their diet, but peninsular Florida contains numerous common *Quercus* species (Brockman, 1968). Many of these grow to tree stature in extensive forests. Yet Scrub Jays are scarcely seen even as accidentals outside of the open scrub. In contrast, western populations regularly occur in oak habitats of many sorts.

The possibility that Blue Jays competitively enforce habitat specificity upon Scrub Jays in Florida is an attractive, if unproven, explanation. Blue Jays are skilled nest predators, and their superior aerial agility allows them to dominate Florida Scrub Jays in habitats containing tall trees, even oaks (personal observation). Several Scrub Jay territories existing in scrub containing numerous tall pines near the study area failed to produce enough to recruit new territory holders, and the territories disappeared. In these areas we have seen Scrub Jays regularly harassed by Blue Jays. As show above (table 3.1), productivity in these environments is much lower than in the open scrub. Although many factors could be producing this result, we suspect that various aspects of coexistence with Blue Jays in the taller habitats of Florida figure heavily in the Scrub Jays' lack of success there.

CHAPTER FOUR

The Pair Bond

SYNOPSIS

The fundamental breeding unit in Florida Scrub Jays is a permanently bonded monogamous pair, which retains ownership and sole breeding privileges in one territory year after year. Pair bonds usually terminate only through death of one member. "Divorce" rate is about 5 percent; divorce occurs primarily when one member becomes temporarily or permanently incapacitated and is replaced. One instance of bigamy has occurred. Its detrimental effects on health and breeding success of the individuals show that monogamy, including sexual division of labor during nesting, has been strongly selected in this population. Delayed breeding is the rule in the Florida Scrub Jay. Pairing and first breeding occur after a minimum of one full year as a helper. Males may remain as helpers up to six years after fledging, while females typically disperse and pair after one or two years of helping. Widowed breeders usually remain on their territory and pair with an immigrant replacement, even when offspring of breeding age are present. Widowed females show a slightly greater tendency to shift territories than do widowed males. Close inbreeding is extremely rare, appears to be actively avoided, and occurs through accidental dispersal by one jay into the territory of an unfamiliar relative. Only rarely do helpers join a pair other than their parents, but some of these "adoptees" eventually become breeders in the new territory. The importance of the pair bond and its continuity through the years show Florida Scrub Jays to be relatively simple, perhaps primitive, in their degree of social cooperation.

The fundamental unit in the Florida Scrub Jay breeding system is a monogamous pair. Every square meter of scrub is occupied by a territory, and every territory contains a mated pair of jays. Working together, members of a pair defend territory boundaries all year; they build nests and perform most or all care of offspring in the spring, and they retain ownership of the same piece of land year after year while each year's offspring mature into helpers and eventually disperse. We followed 280 seasonal breeding attempts over 11 breeding seasons, in which only one instance of polygamy occurred. With rare exceptions, pair bonds terminate only when one member of the pair dies. "Divorce" is rare and seems to occur only when one member of the pair no longer can perform its functions. In this chapter we examine aspects of the pairing process such as mate fidelity, ages at first breeding, fates of widowed breeders, and incest avoidance. Further patterns relating to the initiation of pairs and to dispersal are discussed in chapters five and seven.

4.1. EVIDENCE FOR MONOGAMY

All of our evidence for monogamy is circumstantial, principally because copulation appears to be purposely concealed. It occurs either on the ground amidst dense cover or directly on the nest during the brief egg-laying period. However, closely observing hundreds of pairs literally for thousands of hours convinces us that monogamy is the rule in the Florida Scrub Jay and that the pairs we see tending clutches produced the eggs and sperm in those clutches.

That the clutches contain eggs only of the paired female is supported by the following: (1) we have never found eggs deposited in a nest at a rate faster than one female produces them (one per day), (2) nor even at a rate slower than one per day; (3) we have never encountered an abnormally large clutch (i.e., larger than five eggs); (4) not once have we observed a female other than the breeder even near the nest, much less on it, during the egg-laying period.

Because families include only one incubating female (see 4.3. "Bigamy" for the single exception), and apparently all the eggs she incubates are hers, only the source of sperm is open to question in testing for the existence of monogamy and the absence of cuckoldry. Possible alternative sources of sperm are males from other families and male helpers from the female's own family. Regardless of the source, we exclude forced copulation from consideration because Scrub Jay females are capable of escaping such rape.

The few copulations we have seen (N = about 20) always have been between members of a pair, and always out of sight of all other jays. Because we have learned the circumstances and behavior patterns that precede copulation between members of a pair, we easily could increase our sample of these observations. Instead, we choose to concentrate on following the behavior of intruders, which include adults of both sexes. Not only have we never seen copulation between unpaired jays, but we have never even seen courtship behavior between a female and any jay other than her mate. Copulation does not occur without a lengthy and ritualized courtship process (samples of which we frequently do witness), so we feel certain that extrapair copulation is exceedingly rare, if it occurs at all.

Strict territoriality in an open habitat in which family members tend to stay together makes fertilization by nonfamily jays improbable. The normal reaction by a resident female to an intruder is a scold or hiccup-call if the trespasser is a male, and a chase if it is a female. Some resident

56

males are slow to chase trespassing females, but the resident female is not. Fertilization by a helper male from within the family also is unlikely because of the strict dominance system (Woolfenden and Fitzpatrick, 1977), and because the breeding male nearly always remains in the close company of his mate, especially during the nesting season. The normal reaction of a female breeder to unusually close approach by a male helper is mild scolding and defensive posturing. The call immediately summons her mate. Especially during early breeding season, the male breeder vigorously chases away the helper in these instances.

Normally only immature jays are allowed to stay very long in a neighbor's territory (see, for example, 7.6. "Fledgling Wandering"). Persistent intruders occasionally are accepted into the family, but only as subordinate helpers (see 4.7. "Adopted Helpers"). Even when two or more males vie for the same female and the loser remains in the territory, the loser is subordinate to the male that forms the pair. We have never had evidence that even these likely candidates have copulated with the female breeder.

No evidence for extrapair copulations was obtained during a recent study of extrapair courtship in another corvid, the Black-billed Magpie, *Pica pica* (Buitron, 1983). In contrast to Florida Scrub Jays, magpie males, and especially neighboring breeders, repeatedly visited females on their territories. However, only 15 of 92 visits (male within 5 m of female) progressed to courtship soliciting, and never was food-passing observed. Only once, when both pies were out of sight of the observer, might copulation have occurred. We suspect that strong division of labor during nesting, in the Florida Scrub Jay and perhaps other corvids, would select for strong and faithful pair bonds and against extrapair copulations.

4.2. DURATION OF PAIR BONDS

Because the death rate of breeders is low (see chapter nine), many pairs remain together over several breeding seasons. We observed 93 terminatons of pair bonds, of which 80 (86%) were caused by death of a breeder while only 13 (14%) involved divorce. A more critical measure of pair fidelity, discussed below under "Divorce," shows that of 170 seasonal pairings in which both members were alive the following season, 157 (92.4%) remained paired with each other. Figure 4.1 graphs the annual survivorship of pair bonds in our study population. The constant annual decay rate of .65 shows that, regardless of how long they have been together, members of a pair have a 65 percent chance of being together from one breeding season to the next, a 42 percent chance for two successive seasons, and so on. As expected for a population in which pairs normally remain mated until death of a member, this decay rate in pair survival is almost exactly twice the annual death rate of breeders (see chapter nine).

Divorce

By our definition, divorce occurs when two jays that function as a mated pair during one breeding season are alive but not paired with each other during the next season. Twenty-four individual jays are known to have been involved in divorce, including one male and one female two times each. The total number of known divorces is 13. Dividing this figure by the number of times both members of a pair survive from one breeding season to the next (N = 170) gives a divorce rate of 7.6 percent. As the following case histories suggest, reduced ability to perform certain activities required of breeders may underlie most cases of divorce.

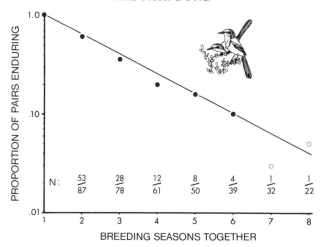

FIGURE 4.1. Annual survivorship of pair bonds. For the first six years at least, regardless of how long a pair has been together, the probability is 65 percent they will be together the next year. Sample sizes (N): denominator = number of pairs we had opportunity to observe together for *x* years; numerator = number of pairs still together *x* years after first forming. Open dots = small samples.

Case histories 1,2,3. In two instances (R–GS♀, B–RS♂) one member of each pair appeared to be sick, and in a third the male definitely was injured (–FRS♂), with a lame wing. Both sick jays appeared healthy again before the next breeding season, but they did not regain their former mates, who by then had re-paired. The injured male was able to defend only a small territory (Appendix E, STIL). His mate divorced him after one breeding season and paired in 1974 with a healthy male in an adjacent territory (Appendix F, PLOT).

Case history 4. In 1973 a male breeder (L–SO♂) lost part of his territory (Appendix E, ROSE), his nestlings, and his mate to a large neighboring family (Y–OS♂ breeder, COPS) that was expanding its territory. The invaders saw but ignored the nest and nestlings, who rapidly lost weight and were near death when taken into the laboratory. The divorced breeder (L–SO♂) was excluded from much of his former territory, and a helper male from the expanding family (BR–S♂) paired with the female (further details are given in case history 29; figure 7.4).

Case history 5. In 1977 a female breeder (F–AS♀) deserted a territory (Appendix I, SBAY) she occupied with her mate and an unrelated female "helper" (YSP–♀). At the time of pair formation, it appeared to us that both of these females had tried to pair with the resident male. The female breeder began an unsuccessful nesting attempt, which appeared to be disrupted by the helper. The following season the former "helper" paired with the male after his first mate departed and paired with a novice breeder elsewhere (Appendix J, POLE).

Case history 6. In August 1973 a female (–AYS♀) became divorced from her mate of two consecutive years (LL–S♂, Appendixes D, E, PLOT) and returned to her natal territory (Appendix F, BIGL), where she helped for five more years (1974–1978) before disappearing. The female mentioned in case history 4, who divorced the lame male, paired with LL–S♂ and may have perpetrated this divorce. During several of the helping years –AYS♀ was absent from home in early spring, suggesting she made a number of dispersal forays, from which she returned home unsuccessful. No other female known to us has helped beyond age 4 years.

Case history 7. A female of the 1969 year class (AA–S♀) dispersed and bred in 1971 with a male (B–WS♂) of unknown history. They had a male helper (R–FS♂), also of unknown history. The male breeder disappeared in February 1972, and the female moved to an adjacent area where she paired and bred in 1972 with a presumed novice (F–GS♂). This male, of unknown age, helped in 1971. In March 1972, after an absence of one month, her first mate reappeared, with a new female. This new pair bred in 1972 with the same male (R–FS♂) as a helper. In August 1972 the once divorced female (AA–S♀) divorced again, and *returned to her original breeding territory*. She resided here with her first mate (B–WS♂), his current mate, and the male who had helped both females. In December 1972 the resident male breeder died, in January his existing mate disappeared, and by February the twice divorced female had paired with her former helper. They bred here (Appendixes E, F, WCUL) in 1973 and 1974, when he died. We suspect the first mate of this twice divorced female was incapacitated in

60

February 1972, and that this perpetrated her first divorce. Possibly both males (B–WS♂, R–FS♂) courted her when she first moved into the territory, which could account for her pairing with a former helper.

Further evidence supporting our opinions regarding causes of divorce comes from an unusually informative case history reported previously (Woolfenden, 1976b). In brief, the events were as follows. After breeding successfully in 1972, a male (L–GS♂) began prebasic molt in June, the usual time; however the quills of all of his new feathers aborted early in development. Although otherwise apparently healthy, the jay became flightless. He could not defend his territory (Appendix D, SORR), and by July he had lost his mate to another male who moved in. At this stage we captured the flightless male for study in the laboratory. This history is not tallied as a divorce because the male certainly would have died before the next breeding season. It does indicate, however, that mate replacement can occur before the jay being replaced has died. Such an event, if followed by recovery of the afflicted jay, becomes tallied as a divorce. Thus, Florida Scrub Jays are mated till death *or affliction* do them part.

4.3. BIGAMY

One case of a male with two mates has occurred during the study. The circumstances leading to this event parallel those thought to account for many divorces, namely physical impairment of one member of the breeding pair. As the details are given elsewhere (Woolfenden, 1976b), here we only outline the case. The singularity of the event and its potential evolutionary implications cause us to isolate it as a noteworthy case history.

61

Case history 8. The original mate (W–SY♀) of a resident male (Y–BS♂) appeared to be sick; she remained hidden in the oak shrubs instead of remaining close to her mate as she had done during previous months. She did not participate in territorial defense. A second female (–FGS♀) moved into the territory (Appendix F, YTRE) and courted with the male breeder. Later the original female appeared to regain her health, as she now came out of hiding. However, in direct encounters she was now subordinate to the new female. Both females nested in 1974, but only the second nest of the dominant female produced young. The male would not feed the nestlings; instead he delivered the food we gave him to the subordinate female. In attempting to raise her young the dominant female failed to obtain sufficient food. The young were grossly underweight and died at fledging. The female parent, also not fed by the male, lost about 9 percent of her normal body weight during nesting. She regained her weight shortly thereafter.

In Florida Scrub Jays breeding females incubate and brood while their mates deliver most of the food needed by both her and the young (DeGange, 1976; Stallcup and Woolfenden, 1978). The case history of bigamy shows that females strongly adhere to their role even in the absence of a functioning mate, and will do so at the expense of their own health. This is circumstantial evidence that division of parental labor between the sexes has become fixed, possibly even genetically so, in the Florida Scrub Jay. Such fixed division of labor supports the widely held notion (Lack, 1968) that monogamy evolves as a means of providing sufficient parental care to raise the maximum possible number of offspring. Polygamy of any sort is now all but prohibited by the differing parental roles of each member of the pair.

4.4. AGE AT FIRST BREEDING

Ages of Pairing

The age at which young animals first pair and breed is a demographic parameter of great consequence in any overall life history strategy (Cole, 1954; Lewontin, 1965; Horn, 1978). Some degree of delayed breeding characterizes virtually all group-breeding birds (Brown, 1978a), and the Florida Scrub Jay is no exception. More interesting still, the data discussed below show that in our population males and females exhibit some important differences at this stage of their life histories. In chapters five and seven we show how these differences relate to skewed sex ratios, different mortality rates, and different dispersal behavior among helpers.

The exact age at which jays first *paired* and obtained a territory or established a new one is known for 51 males and 40 females (table 4.1, figure 4.2). We add to the sample 5 males for whom we could establish only minimum age at pairing. Age of first pairing usually is equivalent to the age of first breeding, although exceptions have occurred: 4 known-age males and their mates showed no signs of nesting during the first breeding season in which they were paired on a territory.

Nearly all Florida Scrub Jays remain as helpers in their natal territories through their first breeding season posthatching. However, given proper social conditions, they are probably *capable* of breeding at this age. Three jays, a male and two females, have formed pair bonds at age 1 year. All three built nests with their respective mates, and the male and one of the females fledged young their first season.

Pairing is most common at or near age 2 years (figure 4.2): of the two-year-olds we followed, 47 percent of the males and 58 percent of the females first paired at this age. Twelve of the 49 males (24%) are known to have paired after helping for at least 3 years, including one male of unknown age who first paired at a minimum age of 7 years (G–YS♂). In contrast, only one of 40 females (2.5%) has

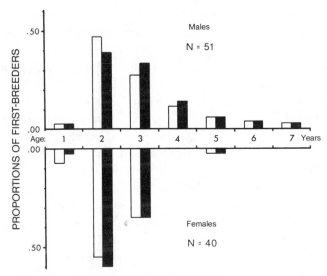

FIGURE 4.2. Age of first pairing (open bars) and first breeding (closed bars) for male and female Florida Scrub Jays. Sample sizes (N) for females are smaller because more of them disperse to become breeders outside the study tract.

delayed pairing beyond age 3 years. The history of this female suggests that even this much delay is exceptional:

> *Case history* 9. One female (S–PW♀) "helped" for 4 years, then dispersed and paired at age 5. However, at age 3 she dispersed and consorted with a helper male in a corner of his natal territory (Appendix H, XRDS). They built a nest and she laid three eggs, which they subsequently abandoned. Both jays rejoined their respective families as helpers. As neither jay was subjected to the rigors of breeding for more than about 6 weeks, we exclude this case from our analyses of pair bonds and ages at first breeding even though they were paired briefly.

Duration as Helpers

The number of years jays have remained as helpers before dispersing and pairing, or dying, is shown in table 4.2.

TABLE 4.1. Age at First Breeding for Florida Scrub Jays

Age in Years	Males[a] (N = 51)		Females (N = 40)	
	No.	Proportion	No.	Proportion
1	1	.02	2	.05
2	24	.47 (.39)	23	.58
3	14	.28 (.33)	14	.35
4	6	.12 (.14)	0	.00
5	3	.06	1	.03
6	2	.04	0	.00
7	1	.02	0	.00

NOTE: Includes 1971–1979 new breeders.
[a] Four of the 24 males tallied under age 2 years paired at this age, but did not nest until ages 3 years (N = 3) and 4 years (N = 1). Proportions of first breeders at ages 2–4 years, using first nesting as the criterion, are shown in parentheses. Ages were estimated for 6 (other) males banded early in the study.

These data include samples ranging from 63 jays (32 males, 31 females) banded early enough to have been observed helping for as many as seven years to 186 jays (87 males, 99 females) that we observed for at least one year. The results (figure 4.3) parallel those described earlier (Woolfenden and Fitzpatrick, 1978a) and support our conclusions regarding age at first pairing: females pair significantly earlier in their lives than do males. The differences between the proportions of females *no longer helping* at age 2 (.69, table 4.2) and the proportion of female *new breeders* at age 2 (.58 + .05 = .63, table 4.1), a difference not found among males (both .49), is important. This difference results from increased mortality among young female helpers and is discussed more fully in chapter seven.

Figure 4.3 demonstrates a fundamental difference in dispersal characteristics between males and females. The following observation, and its implications, will be discussed more fully in chapters seven and ten, but it warrants brief introduction here. The data on "survivorship as a helper"

TABLE 4.2. Duration of Florida Scrub Jays as Nonbreeding Individuals in the Population

	Years as Nonbreeder						
	1	2	3	4	5	6	7
Females							
Number observed	97	27	4	2	0	0	0
Potential sample[a]	99[b]	86	72	65	46	42	31
Proportion	.980	.314	.056	.031	.000	.000	.000
Males							
Number observed	86	36	16	6	2	1	0
Potential sample[a]	87[b]	71	61	57	46	35	32
Proportion	.989	.507	.262	.105	.043	.029	.000

NOTE: Includes a few nonbreeders who associated with pairs outside their natal territory (see 4.7. "Adopted Helpers").

[a] Each sample represents the number of yearlings from those year classes that potentially could have been observed helping for n years after fledging (see 4.4. "Age at First Breeding" for detailed explanation of this calculation procedure).

[b] 186 yearlings include 159 (76♂♂, 83♀♀) of known age and parentage banded as nestlings; 13 (6♂♂, 7♀♀) assumed to be yearlings when banded as "unknown-age" helpers in 1969 and 1970; and 14 (5♂♂, 9♀♀) banded as fledglings, who later immigrated into the study tract.

are plotted semilogarithmically in figure 4.3. Thus a straight descending line (or negative exponential decay) would reflect an annual probability of departure from the territory (either through death or dispersal) independent of time already spent at home. Such a curve does not seem to describe females (who, indeed, disappear entirely as helpers beyond age 4 years). However, the *lower annual probability of departure of males shows no age-dependent increases.* This pattern is consistent with a strategy of waiting in the natal territory until (1) a breeding opportunity arises nearby, an event that would have roughly constant annual probability of occurrence, or (2) death, assuming it also follows a constant, age-independent probability (see chapter nine). This

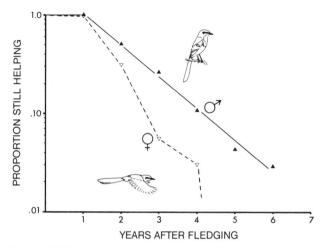

FIGURE 4.3. Duration of jays as helpers, plotted semilogarithmically. Disappearance of males is roughly age independent (straight line). Female disappearances increase with age (curving dashed line).

aspect of male dispersal strategy makes intuitive sense, given the opportunities the jays can create for inheriting breeding space without departing (Woolfenden and Fitzpatrick, 1978a; see chapters seven and ten). We will refer to this result in the process of summarizing the optimal strategies for non-breeders in chapter ten.

4.5. FATES OF WIDOWED JAYS

Widowed jays, as well as helpers, are a source of individuals from which new pairs are formed. Usually a widowed jay, regardless of sex, remains on its territory, and a disperser moves in to attempt pair formation. Deserting the territory is rare for either sex and is especially rare for males.

Only 2 of 36 widowed males (6%) moved to pair in an adjacent territory. In each case the male paired with an experienced, widowed female. Because female jays are dominated by males, including male helpers (Woolfenden and Fitzpatrick, 1977), a female who has lost her mate is in potential conflict with any resident male helpers for breeding space. Perhaps for this reason more female breeders move following death of their mates than do males, although the difference is not statistically significant with our small samples. Nine of 40 widowed females (23%) subsequently disappeared (N = 4) or moved one or two territories away (N = 5) for re-pairing. In one instance a widowed female seemed to be excluded from the territory by her son and two stepsons. She lived peripherally to the group for a few months before disappearing. However, most females (77%), including many (N = 10) with older male sons or stepsons present, remain home and pair with an incoming replacement male, who quickly establishes dominance over the helpers through aggressive encounters. Only 3 of the 9 departing widowed females had older male helpers, and in two cases these helpers actually departed with their mothers.

The slightly greater tendency for breeding females to depart following mate loss probably is a simple result of the females' greater tendency for dispersal (see chapter seven). Given that male helpers clearly dominate their mothers (Woolfenden and Fitzpatrick, 1977), it remains unclear to us why more widowed females are not driven from their breeding territories prior to the gradual process of their forming a new pair bond.

4.6. CLOSE INBREEDING

The forming of a pair between close relatives rarely occurs in Florida Scrub Jays. However, we suspect these jays are poor at identifying genetic relatives. Instead, close in-breeding is rare probably because behavioral traits exist whereby a jay refuses to pair with another member of the functional cooperative-breeding group in which it was raised. Such an "incest taboo," combined with sexual differences in dispersal distances and a low reproductive rate, could cause close inbreeding to be a rare event. Incest avoidance has been shown among highly social black-tailed prairie dogs (*Cynomys ludovicianus*) (Hoogland, 1982) and Cape hunting dogs (*Lycaon pictus*) (Malcolm and Marten, 1982), and probably holds for most, or all, cooperative-breeding birds (e.g., Acorn Woodpeckers: Koenig and Pitelka, 1979).

We know of five exceptions to the generalization that close genetic relatives do not form pairs. Four of these exceptions involved seven jays that never functioned as members of the same breeding group. All had descended from the same extraordinarily prolific breeder, whose history is outlined in Appendix L. The improbable event of long-term success at producing breeders, coupled with the limited dispersal characteristic of the population, explains these matings, the genealogy of which is presented in figure 4.4.

The outstanding exception, in which a son paired with his mother, appears to have occurred because of the infirmity and eventual death of the breeding male (WSW–♂), which left his mate, their helper son, and a fledgling son as the only occupants of the territory (Appendix K, NBEN).

Case history 10. During the winter of 1978–79 a male breeder (WSW–♂) developed a lame leg, which reduced his effectiveness at defending his territory (Appendix J, NBEN). Territorial defense was increased by his son (SPG–♂) and the mate of the cripple. On several occasions we witnessed the son drive off foreign males who were visiting the territory and apparently trying to pair with the female. The crippled breeder died in

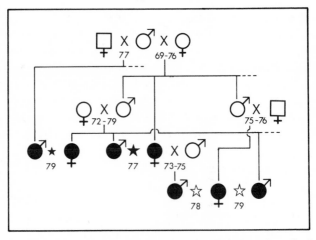

FIGURE 4.4. Genealogies of seven jays (closed circles) who paired (stars) with close relatives. Dates below pairs show the years of their duration between 1969 and 1979. One female paired during successive years (open stars) with two close relatives, and one immigrant female (square) paired first with a son and later with the son's father. Dashed lines indicate that more surviving offspring were produced by a pair than are shown here.

December, no outsider succeeded in moving in, and late in the breeding season the mother and her son formed a pair and produced fertile eggs (Appendix K, NBEN). The female died during incubation and was replaced immediately by a dispersing female who had been visiting the territory prior to the breeding female's death. The new pair also nested, albeit unsuccessfully, the same year (Appendix K, NBEN). A resident younger male ($-SRY\,\delta$) was usually present but seemed not to be involved in the interactions.

4.7. ADOPTED HELPERS

Florida Scrub Jays normally exist only as members of a breeding group. Increased survival certainly is one of the advantages of this habit (see chapter nine). If both members of a pair die between successive breeding seasons, any helpers that remain either continue to defend the territory while the dominant male helper gains a mate from outside the group, or they leave and attempt to join other established groups. Behavioral details regarding outsiders joining established groups are given in chapter five (section 5.4. "Whom Helpers Help"; see also case history 15).

Shifts of helper jays from one family to another may occur even without family disintegration. Furthermore, the fact that some of these jays have become breeders within the group to which they transferred suggests an advantage of joining a group of jays that contains no relatives. Case history 5 and case history 8 (bigamy) present two examples of a jay joining an established group and eventually becoming a breeder within it. Two more jays have become breeders in families they joined as "helpers." These cases clearly show that instances of adoption can be viewed as a dispersal strategy by nonbreeders.

Case history 11. In October 1978 two siblings, hatched that year (including −SAY♂), departed from their parents (Appendix J, NRID) and joined a family consisting of an experienced pair (−ALS♂, L−SR♀), their oldest helper son (SBW−♂), an immigrant female who eventually paired with this son, and four 1978 young (two males, two females; Appendix J, SAND). The oldest son and the immigrant female budded off a territory (Appendix K, SSAN) by 1979. During the ensuing 1979 nesting season, the four remaining offspring of the experienced pair and the two adopted helpers attended two consecutive nests. That all was not stable within this extraordinarily large group is suggested by two facts: (1) one of two tiny young disappeared from the first nest, and (2) one of the two young in the second nest was grossly underweight (21.0 g vs 40.2 g at day 11) and disappeared before fledging.

During the second nesting attempt, one adopted jay disappeared; the other (−SAY♂) remained but continued to be dominated strongly by the two resident helper males. Between July and August the male breeder (−ALS♂) died, and by September

71

the *adopted male had paired with the widow*. Now he clearly dominated the two resident sons who had dominated him for an entire year! These males disappeared within a few months thereafter: one was seen in the budded territory of his older brother before disappearing; the other son was attempting to establish a territory nearby when he disappeared. The two females disappeared, one in July, the other in August, the usual months of female helper dispersal. The "adopted" helper, now paired with the widowed, experienced female, nested without helpers starting in 1980.

Case history 12. In April 1979 a female of the 1978 year class (–SAB♀) began wandering from her natal territory (Appendix J, POLE) where her father and stepmother still resided. Possibly she was driven from home by the female breeder, who dominated her strongly. By July she was residing with a pair and their daughter of the 1978 year class in a territory nearby (Appendix K, BIGL). Between July and August the resident female (R–WS♀), who was at least 12 years old, died. In August *the adopted female formed a pair bond with the widowed male*, and by September his daughter had disappeared. The new pair first bred in 1980.

Case history 11 raises the question of why the two fledglings departed from home. In contrast to the jays in case history 12, we detected no instability in their family. Fledgling wandering (see chapter seven) is a possible explanation. The group of four young fledglings, an older male helper, and an immigrant female may have appeared to be a wandering troop to these two young jays. Perhaps after remaining with them for a while, they could not find their way home, even though it was only three territories and less than one kilometer away. The other possibility, which also might apply to the wanderer in case history 12, is that the immigrants, who later became helpers and eventually breeders, detected some weakness in the pair bond between the breeders in the groups they joined. Another enigma, particularly true for the male wanderer who became the breeder in case history 11, is how a subordinate helper eventually gains dominance over his former dominants. The activities of the widowed jay may help this change to come about, but clearly the sudden attainment of breeding status by this "adopted" jay completely changed the social relationships within the group. It is difficult to avoid concluding that the success of the foreign jay at obtaining breeding status in the new group passively or actively caused the two resident males to leave.

4.8. ADVANTAGES OF MATE FIDELITY

The single instance of bigamy showed that a male was incapable of simultaneously providing food to two incubating females. One female was all but abandoned when the male's original mate regained her health and began to nest. It was probably not coincidence that the male chose to provide for his previous mate. The pairing process between two unfamiliar jays often takes months to complete, and it may take several breeding seasons before a new pair can efficiently work together to maximize their chances of raising a successful brood. For this reason it becomes advantageous for breeding individuals, particularly those residing permanently on a territory, to retain the same mate as long as possible. We have evidence that this is the case for Florida Scrub Jays.

In table 4.3 we show the average reproductive success of pairs whose bond has endured from one to eight breeding seasons. Each mate replacement represents a new pair bond in our analysis. As measured both by the percentage of fledging success and by average fledgling production per breeding season, newly formed pairs show significantly inferior reproductive success compared to enduring pairs,

73

TABLE 4.3. Fledgling Productivity as Related to Duration of the Pair Bond in Florida Scrub Jays

	Duration of Pair Bond (Years)							
	1	2	3	4	5	6	7	8
Fledging success (%)[a]	60	77	88	79	73	78	67	100
Fledglings/pair ($\bar{x}$)	1.5	2.3	2.4	2.4	2.3	2.4	3.0	3.5
S.D.	1.5	2.6	1.5	1.6	1.5	1.5	—	—
N (pairs)	105	52	25	19	15	9	3	2

[a] Fledging success = percent of nests fledging at least one young.

especially if they lack helpers (figure 4.5). Table 4.3 shows that if Florida Scrub Jays were to obtain new mates every year, their annual (hence lifetime) reproductive output would be only about two-thirds that actually achieved by retaining the same mate.

The greater reproductive output of enduring pairs results from a variety of changes as members of a pair grow familiar with one another. New pairs frequently show lingering courtship behavior well into the breeding season, and their territorial boundaries often remain in some dispute. For these and other reasons, their initial nest attempts often are delayed relative to those of experienced, enduring pairs. As shown in chapter eight, early seasonal nest attempts show far greater success than later ones. Clutch sizes are slightly smaller among new pairs, even where the female has nested (with a different mate) in previous years. Finally, new pairs often, but not always, include jays who are true novices entering the breeding ranks for the first time. As in many bird species, this alone seems to reduce the effectiveness of the pair at bringing off a successful nest. Novice breeders may even abandon nests with contents midway through the season. These patterns are further elaborated in chapter eight.

74

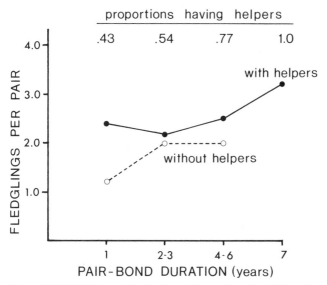

FIGURE 4.5. Fledgling production in relation to duration of the pair bond for pairs without helpers (dashed line) and pairs with helpers (solid line). As pairs endure, their probability of having helpers increases (top).

Early in their history as a pair, if two jays remain bonded from one breeding season to the next, their success at fledging young increases (figure 4.5, dashed line). Increased fledging success increases the probability of having helpers next time around (top of figure 4.5), and the existence of helpers increases fledging success (see chapter eight). Breeder deaths and divorces, which force one or both members of a pair to form new bonds, for various reasons (e.g., section 4.5) often result in the loss of helpers. Both of these factors reduce fledging success (figure 4.5). Thus, in Florida Scrub Jays, mate fidelity improves reproductive success not just through shared experience, but also in improving the probabilities that the pair will be assisted by helpers.

Probably in every aspect of routine life as breeders, jays

who have worked together as a bonded pair in previous years operate more efficiently with one another and their surroundings than do jays who are learning their new home or companion.

4.9. CONCLUSIONS

In the Florida Scrub Jay an individual must become a member of a monogamous pair in order to leave any offspring of its own. In later chapters we discuss the importance of gains in fitness achieved through helping. It is obvious, however, that any such indirect fitness enhancement (i.e., other than direct reproduction) must be evaluated relative to the strategy of obtaining a "lifelong" mate and producing one's own offspring.

In its simple, one male–one female mating system the Florida Scrub Jay resembles the vast majority of birds (Lack, 1968). The monogamous pair is believed prevalent primarily because extensive care usually is required for successful rearing of young, and substantially more young can be reared by two parents than by one (Emlen and Oring, 1977). In our population, this notion is supported by the enormous time and energy expended by both parents in raising a brood (DeGange, 1976; Stallcup and Woolfenden, 1978), and the extent to which division of this labor has become fixed (see 4.3; Power, 1980). We would add that a positive feedback system may exist that reinforces permanent monogamy in species such as the Florida Scrub Jay and that may prevent any occasional, exploratory, or facultative polygamy from ever succeeding. This idea, which is supported by our evidence that enduring pairs show greater fledgling productivity than do newly bonded ones (table 4.3), is suggested in an observation made by Emlen and Oring (1977, p. 217):

Birds breeding with former mates . . . [show] a demonstrable increase in reproductive success. The longer the period of mate fidelity, the more the future physical condition of a mate becomes of importance to its partner. It then becomes adaptive to equalize the energetic burden of reproduction and to share in parental care.

As these shared interests become great, certain energy-saving behavior patterns, such as division of parental labor between sexes, should be favored. The fixation of any such division then actually prohibits opportunities for occasional, successful polygamy because neither sex can function effectively as a parent without the other. The fortuitous instance of bigamy in our population was one case in point. Permanent monogamy is thereby reinforced as an evolutionarily highly stable strategy once it becomes favored at all.

It is difficult to generalize on the importance of the pair bond in cooperative-breeding birds, primarily because too little is known about most of the interesting, potentially more complex species. It is clear that Florida Scrub Jays represent the simplest possible extreme with regard to mating systems among cooperative breeders. With few exceptions (see chapter seven, section 7.8) only a single pair resides in each territory. The two mated individuals are the only birds contributing to the parentage of offspring raised therein. This common pattern is reported for a wide variety of cooperative breeders (Brown, 1978a). Examples include the Superb Blue Wren (Rowley, 1965); Jungle Babbler (Gaston, 1977), Common Babbler (Gaston, 1978b), and Hall's and Grey-crowned babblers (Brown, 1978a; Brown and Balda, 1977); White-bearded Flycatcher (Thomas, 1979); Puerto Rican Tody (Kepler, 1977); Kookaburra (Parry, 1973; Morton and Parry, 1974); Red-cockaded Woodpecker (Ligon, 1970); and several other species of jays (Crossin,

1967; Alvarez, 1975). Brown (1974, 1978a) recognizes this as the simplest, and probably the most evolutionarily primitive, cooperative-breeding system.

Recognizable, monogamous pair bonds remain apparent among some species having multiple nests within group territories (e.g., Mexican Jay: Brown, 1970, 1972; other jays: Hardy, 1976) or even more complex systems of cooperative breeding (e.g., Green Woodhoopoe: Ligon and Ligon, 1978, 1979; Groove-billed Ani: Vehrencamp, 1978). Even along an "alternate route to sociality" (Brown, 1974), *colonial* group-breeders show clear pair-bond relationships and even some permanence to these bonds between breeding seasons (e.g., Pied Kingfisher: Reyer, 1980; White-fronted Bee-eater: Emlen, 1978; other bee-eaters: Fry, 1972a, 1972b, 1975).

Several species of cooperative-breeding birds are reported to be polygamous. Examples include Galapagos Hawk (de Vries, 1975; Faaborg et al., 1980; Faaborg and Patterson, 1981), Harris Hawk (Mader, 1979), Tasmanian Native Hen (Ridpath, 1972; Maynard Smith and Ridpath, 1972), Pukeko (Craig, 1976, 1977, 1979, 1980a, 1980b), and especially the Acorn Woodpecker, where polyandrous mate-guarding has been documented (Mumme et al., 1983). Recently, even electrophoretic evidence for a case of multiple paternity has been obtained (Joste, 1983). Each of these examples involves females consorted by multiple males, and the term *cooperative polyandry* was coined to describe such cases (Faaborg and Patterson, 1981).

Unequivocal evidence supporting persistent monogamy or polygamy can come only from genetic data, and it is universally lacking for cooperative-breeding birds. We have detailed evidence supporting monogamy in the Florida Scrub Jay, but still it is circumstantial and therefore open to question. We suggest, however, that good reasons exist for being

78

skeptical of assumed polyandry, even in the cases mentioned above. First, cooperative breeding almost universally involves auxiliary males more than females (Brown, 1978a). Furthermore, it is substantially more difficult to prove or disprove multiple paternity than multiple maternity. Therefore, possible cases of polyandry are both more numerous and more difficult to eliminate unequivocally.

Second, monogamy typifies closely related species of virtually all cooperative breeders, including the list of possible polyandrous ones just mentioned. The possibility exists that pair bonds, albeit subtle ones, persist in many of the equivocal cases, or that seemingly promiscuous or polyandrous groups actually are intermediate, competitive conditions during the establishment of a pair. For example, Craig (1980a) clearly indicates that Pukekos seem to strive to become permanently monogamous.

Third, a paired male can gain from allowing other males to copulate with his mate during times when her eggs cannot be fertilized, because this increases the other males' likelihood of remaining to help rear the brood. Such a system, while it may be nearly behaviorally identical to the ideal version modeled by Faaborg and Patterson (1981), and may indeed be cooperative, is not polyandrous in the genetical sense. We conclude this chapter by reminding ourselves that we must firmly establish which pair-bond system our animals practice before we can draw safe conclusions about its relationship to cooperative breeding.

CHAPTER FIVE

Helpers

SYNOPSIS

Helpers are nonbreeders. They assist the breeding pair in all territorial and breeding activities except nest construction, egg-laying, and incubation. Only about half of the breeding pairs are attended by helpers each season. Pairs may have up to six helpers, but most have from zero to two. Average group size is three jays, and only 10 percent of the breeding pairs have three or more helpers. The proportion of pairs with helpers and number of helpers per pair both vary directly with average reproductive output during the preceding year. Most helpers are yearlings that remain home and help their parents. Helping beyond the age of two years is infrequent, especially among females, but a few males have helped up to six years before breeding. The sex ratio among yearling helpers is even; older helpers are mostly males. Two-thirds of the helpers assist both full parents, 90 percent assist at least one parent. Jays continue as helpers when their parents are replaced by unrelated breeders. When families break up through loss of both breeders, helpers usually succeed in joining unrelated families as subordinate helpers. A dominance hierarchy exists in all families: males dominate females, breeders dominate helpers, older helpers dominate younger ones, and helpers of like sex and age show linear dominance-subordinance relationships. This ranking among helpers appears to dictate the sequence by which helpers become breeders.

In comparison with other cooperative-breeding bird species, the Florida Scrub Jay appears to exhibit a primitive stage of helping behavior. Site tenacity beyond the normal age of dispersal seems to be the key event in the evolution of this helper system.

Virtually all helpers are prebreeders, and most are the offspring of the pair in whose territory they reside. As jays may remain nonbreeders for several years (table 4.2), helpers inevitably accumulate in certain territories. Because they tend to remain in the natal territory even after replacement of one or both parents, helpers sometimes do aid nonrelatives. If the group in which a helper resides disintegrates (usually through breeder deaths), young helpers may attempt to join another group, sometimes successfully. These "adopted" jays become an additional source of helpers who are not close relatives of the breeding pair or any offspring they help.

In this chapter we summarize our information on the numbers, activities, and sex ratios of helpers, and on the genetic affinities between the helpers and the breeders they assist. Intrafamilial social dominance plays an important role as helpers compete for breeding vancancies (Woolfenden and Fitzpatrick, 1977), and our previous data and conclusions on this subject are updated briefly as well. In later chapters we discuss various effects of helpers on territory size (see chapter six) and reproduction (see chapter eight) and their patterns of dispersal into breeding slots (see chapter seven).

5.1. NUMBER OF HELPERS

During the 11 years 1969 to 1979 we followed the activities of numerous pairs through 280 seasonal breeding attempts. Table 5.1 summarizes the frequency of helpers during these years. At least one helper was present for just over half (57%) of the seasonal breeding attempts. Through this period the frequency of pairs with helpers varied from 41 to 80 percent (mean of annual percentages also equals 57%). The average number of helpers per pair has ranged

TABLE 5.1. Number of Helpers per Family, 1969–1979

Breeding Season	Number of Pairs	Number of Helpers							Helpers/Pair
		0	1	2	3	4	5	6	
1969	6	33.3	50.0	16.7	—	—	—	—	0.83
1970	16	50.0	25.0	12.5	12.5	—	—	—	0.88
1971	25	20.0	36.0	28.0	8.0	4.0	4.0	—	1.52
1972	30	50.0	20.0	20.0	10.0	—	—	—	0.90
1973	26	57.7	26.9	11.5	—	3.8	—	—	0.65
1974	29	48.3	31.0	13.8	3.4	—	—	3.4	0.90
1975	27	59.3	18.5	18.5	3.7	—	—	—	0.67
1976	27	29.6	25.9	18.5	14.8	11.1	—	—	1.52
1977	28	42.9	21.4	32.1	—	3.6	—	—	1.00
1978	32	50.0	31.3	3.1	9.4	6.3	—	—	0.91
1979	34	29.4	41.2	17.6	8.8	—	2.9	—	1.18
1969–79	280	43.2	28.6	17.5	6.9	2.9	0.7	0.4	1.00
$\overline{X}$ of $\overline{x}s$	—	42.8	29.7	17.5	6.4	2.6	0.6	0.3	1.00

NOTE: Based on censuses taken early in each breeding season (March–April).

from .65 to 1.52 ($\overline{x}$ = 1.00). Considering only pairs with at least one helper, the average number of helpers per pair ranged from 1.55 to 2.16 ($\overline{x}$ = 1.78). Few pairs have more than three helpers during a breeding season, and no pair has had more than six.

Sizable annual ranges in average number of helpers per pair reflect annual variation in reproductive success within the population during the one or two years preceding each nesting season. This point is demonstrated by figure 5.1, in which average number of helpers is graphed against fledgling productivity the previous year. The solid line shows a linear regression with significant positive slope. The dashed, curved line (fitted by eye) may reflect the real pattern more closely. As shown in chapter nine, fledgling survival to age 1 year (their first season as nest helpers) is positively correlated with fledgling production, resulting in relatively more yearling helpers during any year that follows a very

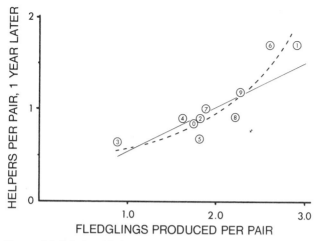

FIGURE 5.1. Relationship between annual fledgling production and number of helpers in population one year later. The circled numbers, 0–9, represent the years 1970–1979. The solid line shows a linear regression through points. The dashed line may reflect the real relationship more closely, because fledgling survival is positively correlated with fledgling production.

successful breeding season. The general relationship shown in figure 5.1 holds because most helpers are yearlings (see table 5.2).

5.2. ACTIVITIES OF HELPERS

Except by their color-rings, helpers are essentially indistinguishable from breeders, in plumage and in daily behavior patterns, during all activities except nesting. All foraging, resting, roosting, territorial, and antipredator behavior qualitatively resembles that of the breeders. Group members spend much of the day in close proximity to one another, and helpers often participate in sentinel activity while the rest of the group forages. Helpers regularly join the breeding pair in territory defense, including active displays, chases, and even fights along territory boundaries.

In this respect helpers appear slightly less aggressive than do breeders (Woolfenden and Fitzpatrick, 1977), but the differences are subtle and quantitative.

Helpers participate in certain nesting activities, but not others. Although they occasionally carry nest material, helpers do not build nests. They do not incubate or brood; these activities are performed by the female breeder only. Rarely, an older helper male feeds the female breeder, but normally she is fed only by her mate. Helpers do feed the nestlings and young fledglings, with males delivering more food than females, and older helpers delivering more than younger ones. Helpers mob predators, and these include nest predators. Some of these nest-related activities, including data on feeding rates of helpers, are discussed in detail in Stallcup and Woolfenden (1978).

5.3. SEX AND AGE OF HELPERS

We determine the sex of young jays through their behavior after reaching independence (chapter two). Often these behavioral signs are not recorded until the fledglings are one year old, when they are helping at the nest for their first time. Therefore the earliest sex ratio we can report is that of yearlings, which is even (81 males: 85 females; table 5.2). If survival over the first twelve months is equal between the sexes, Florida Scrub Jays have an even sex ratio at hatching.

As shown in table 5.2, the helper sex ratio becomes skewed toward males after the first year of life. Considering only the helpers two years old or older, the sex ratio is about 1.6:1 (60 males: 37 females). Among helpers age three years or older, the ratio becomes 3.3:1 (26 males: 8 females). The greater dispersal tendencies among female

TABLE 5.2. Sex Ratios and Age of Florida Scrub Jay Helpers

Breeding Season	Helper Age in Years							Pooled Ratio
	1	2	3	4	5	6	7	
1970	4:7[a]	1:2	—	—	—	—	—	5:9
1971	14:10	4:4	1:1	—	—	—	—	19:15
1972	8:4	7:4	3:0	1:1	—	—	—	19:9
1973	4:2	4:1	3:0	2:0	1:0	—	—	14:3
1974	4:8	3:1	5:0	1:1	1:0	1:0	—	15:10
1975	9:2	0:1	1:0	1:0	1:1	—	—	12:4
1976	10:16	5:1	—	1:0	—	0:1	—	16:18
1977	2:10	6:6	0:1	0:1	—	—	0:1	8:19
1978	10:13	0:5	2:0	—	—	—	—	12:18
1979	16:13	4:4	1:0	—	—	—	—	21:17
Total	81:85	34:29	16:2	6:3	3:1	1:1	0:1	141:122
Ratio	0.95	1.17	8.0	2.0	3.0	1.0	—	1.16

[a] Ratios of known-sex helpers are expressed (males:female) for each age class in which at least one helper existed. Excluded are 14 helpers (10 yearlings, 4 two year olds) whose sex was not positively determined.

helpers (see chapter seven) account for this preponderance of males in the older helper ranks. Only four females account for these eight female helper-seasons (table 5.2); all had atypical histories. Two (–AYS♀, S–PW♀) were known breeders who later returned home to help. They were discussed above (case histories 6,9); here we give information on the other two old female helpers.

> *Case histories* 13,14. In 1976 a female (SP–B♀) paired and probably bred outside the study tract at age 2. In 1977 she returned to join the unrelated family now residing in the territory where she had helped at age 1, and helped there at age 3. Early in the study a female of unknown age (B–YS♀) helped for 4 years (1969–1972). In 1970 she became blind in one eye, which seemed to impair her ability to gain a mate; she disappeared and probably died after helping in 1972.

Few females remain at home beyond age 2, while many males remain much longer. However, it is worth emphasizing here that the death rate among breeders is precisely equal between the sexes (see chapter nine). Hence, ultimately, dispersal-related mortality must account for the skewed sex ratio among older helpers. This too is discussed in chapter nine.

5.4. WHOM HELPERS HELP

We have data on genetic relatedness between helpers and the breeding pairs they helped for 165 individual jays of known age and parentage during 251 helper seasons. These data (table 5.3) show that by far the most common association is that between first-year jays of either sex and their full parents (degree of relatedness to potential nestlings, $r = .50$). Sixty-four percent of all helpers resided with both parents. Moreover, 90 percent of the sample helped at least one parent, including a sizable fraction (24%) who helped a parent and an unrelated replacement (stepparent) following death of one true parent. In most of the remaining cases both parents had died, or peculiarities of new territory establishment led to nonbreeders dwelling with close relatives other than parents (see below).

Because helpers basically represent parent-offspring associations, and the pair bonds of parents persist from year to year with only a 65 percent survival rate (see chapter nine), the average genetic relatedness between helpers and the jays they help raise would be expected to decline with age of the helper. This would be true, however, only if helpers *do not* leave home with greater frequency following the death of a parent, after which their relatedness to potential nestlings would be lowered. We can test the decline in relatedness with the data in table 5.3. Assume that off-

TABLE 5.3. Whom Helpers Help: Family Relationships between Breeders and Their Helpers

Breeders	r	Males	Females	Sex unknown	Total[a]	Percent
Brother × Mother	.75	1	—	—	1	0.3
Father × Mother	.50	69	76	16	161	64.1
Grandfather × Mother	.38	—	1	—	1	0.3
First Cousin × Mother	.35	1	1	—	2	0.8
Brother × First Cousin	.31	—	1	—	1	0.3
Father × Stepmother[b]	.25	21	9	—	30	12.0
Stepfather × Mother	.25	10	17	4	31	12.4
Brother × Nonrelated	.25	2	1	—	3	1.2
"Brother" × Nonrelated[c]	.25	4	—	—	4	1.6
"Uncle" × Grandmother	.19	2	2	—	4	1.6
"Uncle" × Half Sister	.19	—	1	—	1	0.3
Half Brother × Nonrelated	.13	3	—	—	3	1.2
Nonrelated × Nonrelated	—	2	6	1	9	3.6

[a] Based on 165 individual jays of known age and parentage during 251 helper seasons.
[b] Stepparent designation used for nonrelated breeders that paired with a helper's parent.
[c] Quotes indicate probable family relationships.

spring are related to their full parents by a factor of .50 (i.e., no successful close inbreeding in the population) and that immigrant replacements (stepparents) are genetically unrelated to the helper (thus reducing relatedness between helper and potential nestlings by one-half). The predicted equation for average relatedness, r_τ, between helpers of age T years and any nestlings they help raise is:

$$r_\tau = .5 \, (.820)^\tau$$

where r between full sibs is .5 and mean annual survivorship of adult parents is .820 (see chapter nine).

This equation predicts a decline in relatedness shown by

the dashed line in figure 5.2. If helpers leave home signif- icantly more often when their relatedness to nestlings drops (a result that would be consistent with strictly kin-selected helping behavior), the average relatedness actually ob- served between helpers of various ages and the nestlings they help raise would be higher than that predicted by the above equation. Instead, the observed average relatedness in our population declines with age exactly as predicted (figure 5.2). Therefore, within their first three years of life at least, helpers appear to remain regardless of the degree by which they are related to the offspring they assist. This observation is borne out by a large number of individual case histories, too numerous to describe here.

Only 17 jays helped breeding pairs that did not include at least one parent. Five of the 17 helped their uncle during the same season that they were helping their parents, when two brothers nested in the same, jointly defended territory (see section 7.8. "Multipair Territories"). Three others of the 17 helped brothers who bred in the family territory after both parents had died. Two others helped a probable brother after apparently being driven from their natal ter- ritory by an aggressive stepfather (details in Woolfenden, 1975, table 8).

Jays whose own families have broken up, usually through simultaneous deaths of their parents, make concerted ef- forts to join established families. We have watched five jays do this, of which four were birds of the year (see Wool- fenden, 1975, table 9). The process by which jays join an unrelated family is gradual, with frequent begging and other submissive acts preceding acceptance. We interpret these cases as evidence for the importance of Florida Scrub Jays living in a group. In June, 1978, D. Moskovits studied the adoption of two fledglings, providing a case history that typifies the process.

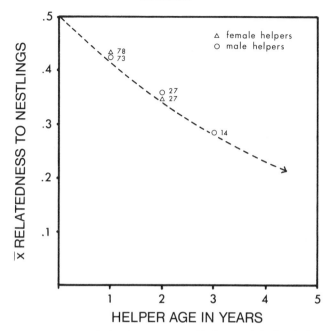

FIGURE 5.2. Correlation between age of helper and decline in its genetic relatedness to the nestlings. The observed average relatedness through time (open triangles and circles) almost exactly fits the predicted decline (dashed line) based on annual mortality of the helpers' true parents.

Case history 15. A family consisting of a breeding pair, a female helper, and one fledgling, age 6 weeks, adopted two fledglings of similar age. On 5 June the family, which at this time included two fledglings, was alone and intact. On 10 June one of their young had died, and two intruder fledglings were present. The two intruders followed the feeders persistently and begged incessantly. Often all three dependent young tumbled over each other as they hounded an adult begging for food. The helper never was seen to feed the intruders, and the male breeder seemed more reluctant than his mate, but within two weeks both breeders fed the intruders often. Initially, choice food (e.g., insects) seemed to be preferentially given to their own, lone young. Peanut bits we provided were

89

doled out less discriminately. By late June the fledglings were finding much food on their own.

We do not know the source of the intruders, although probably they came from a nearby family broken up by death. The family that adopted the two young (Appendix J, CENR) was at the periphery of our study tract. Dependent young who lose their parents and helpers join another family or die. Families with few fledglings of their own may be more susceptible to adopting the young of others.

5.5. INTRAFAMILIAL DOMINANCE

Florida Scrub Jay families function under strict rules of dominance-subordinance (Woolfenden and Fitzpatrick, 1977). Within families males dominate females, breeders dominate helpers, and helpers dominate young fledglings. When two or more male helpers live in the same family, a straight-line hierarchy exists among them, usually with the oldest male dominant. The same pattern, though less conspicuous, appears to hold among female helpers. Earlier we proposed that the intramale dominance system within helpers provides dominant males maximum opportunity to become breeders (Woolfenden and Fitzpatrick, 1977). Previously we had witnessed 14 cases in which a male helper successfully dispersed from a family with two or more older male helpers. In 9 cases the dominant jay paired first. We now add 10 more cases (table 5.4), and in 7 of these again the dominant male was the first to pair. As before (Woolfenden and Fitzpatrick, 1977, table 5), the apparent exceptions include equivocal cases.

Case history 16. Two brothers from the same brood in 1974 helped in their natal territory in 1975 and 1976 (Appendixes G,H, COPS). We recorded numerous interactions between these

TABLE 5.4. Time and Order of Pairing by 17 Males from 8 Families Containing 2 or More Male Helpers of Breeding Age

Male Helper	Year			Dominant First	Mode of Pairing
	Hatched	Helped	Paired		
−BYS>	1970	1971−75	1975 (Nov)	yes	inheritance
SY−P	1974	1975	1976 (Mar)		bud
SW−G>	1974	1975−76	1976 (Aug)	?[a]	bud
SR−G	1974	1975−76	1976 (Jun)		replacement
YGS−>	1971	1972−75	1975 (May)	yes	replacement
SR−Y	1974	1975	D1975 (Aug)		—
SP−R>	1974	1975−76	1976 (Jul)	yes	replacement
SG−R	1974	1975−76	1976 (Aug)		replacement
SY−P>	1974	1975	1976 (Mar)	yes	bud
PSG−	1975	1976−77	D1977 (Aug)		—
RS−P>	1972	1973−76	1976 (Sep)	yes	replacement
WSW−	1975	1976	1977 (Apr)		bud
LS−P>>	1972	1973−75	1976 (Sep)	yes, yes	replacement
FSP−>	1975	1976−78	1978 (Sep)	no[a]	inheritance
ASP−	1975	1976−77	1978 (Apr)		bud
SBG−>	1977	1978−79	1979 (Aug)	no[a]	replacement
SYG−	1977	1978	1979 (Mar)		de novo

NOTE: Spaces in the columns separate helpers of different families; > identifies jays known to dominate their siblings; D indicates disappearance (death or dispersal) from tract before pairing occurred.

[a] Exceptions to the generalization that dominants pair first are discussed in text.

males, and dominance-subordinance was clear (SW−G♂ > SR−G♂). Following successful nesting by the family in May 1976, the subordinate helper (SR−G♂) replaced a dead breeder in an adjacent territory (Appendix H, SWCO) in June 1976. On the June and July censuses the dominant helper (SW−G♂) was away from home at the Main Grounds near the buildings of Archbold Station in an area not defended by any jays. An apparently unpaired female (SY−G♀) also was seen here fre-

91

quently. By August 1976, only two months after the pairing of his subordinate brother and half a year before the next breeding season, the dominant male and this female had paired and come to reside in a corner of his natal territory (Appendix I, NWCO).

Case history 17. During the 1975 breeding season a male (LS–P♂) of the 1972 year class helped raise three fledglings (including FSP–♂ and ASP–♂). The group, which lost the male breeder in July 1975, apparently did not breed in 1976. In July 1976 the two younger helpers moved to an adjacent territory where the male breeder had been killed, and both attempted to pair with the widow. The dominant of the two (FSP–♂ > ASP–♂) fed her most often. In August the older male (LS–P♂) joined them, asserted his previously established dominance, and paired with the widow. The two subordinate males helped the breeders in 1977. The dominant male helper (FSP–♂) occasionally fed the female, a rare event for Florida Scrub Jays. The subordinate helper (ASP–♂) budded off a territory (Appendix J, DOME) in 1978, while the dominant continued to help. Following the 1978 breeding, the male breeder (LS–P♂) died, and the remaining helper paired with the female, another rare event for Florida Scrub Jays. In the interim the male who budded had died.

Case history 18. Two brothers from the same 1977 brood helped in their natal territory in 1978. Numerous interactions showed a clear dominance-subordinance hierarchy (SBG–♂ > SYG–♂). In March 1979 the subordinate male paired with another novice, and together they carved out a small territory *de novo* (Appendix K, JUNK). Two sets of eggs were laid; the second set hatched. In late May the pair deserted these nestlings and returned to their natal territories where they helped care for the existing fledglings. The dominant male (SBG–♂), who had helped at home throughout the 1979 breeding season, replaced a lost breeder in an adjacent territory in August 1979.

As with our earlier analysis (Woolfenden and Fitzpatrick, 1977, p. 7), details of each case history suggest reasons for the exceptions to the general rule that the dominant male helper within a family pairs first. In evaluating the exceptions it is worthwhile to have in mind the ways in which males pair and become breeders. These are by (1) mate replacement, usually away from home; (2) mate replacement at home, which occurs very rarely; (3) territorial budding; or (4) establishing a territory *de novo* between existing territories (see chapter seven for further details). Assuming space for breeding is limited, and space familiar to a potential breeder is optimum, then inheriting the natal territory seems the best option, followed by budding, replacement elsewhere, and last *de novo* establishment.

In case history 16, the dominant helper (SW–G♂), who paired only a short time after his sib, budded a territory, while his subordinate brother replaced a breeder who occupied space adjacent to his natal home. In case history 17, the initial attempts of the dominant brood mate (FSP–♂) to pair with the widow may have worked against his deserting the territory, which eventually he inherited. In case history 18, the earlier breeding of the subordinate (SYG–♂) in a *de novo* territory clearly was a failure, as shown by his desertion and return home to help.

5.6. CONCLUSIONS

Several surveys of birds with helpers have appeared recently. Important examples include four papers organized by Rowley (1976) and more recent reviews by Brown (1978a) and Emlen (1978). Clearly the number and variety of bird species known to have helpers is vast and increasing rapidly. Many unique behavioral characteristics exist in one or a few species, and only a few traits seem common to many. Especially frequent traits, apparently nearly universal among cooperative-breeding jays, are (1) individuals delay breeding beyond their time of physiological maturity, (2) sex ratio is biased toward male helpers, (3) helpers are nonbreeders, and (4) helpers usually assist their parents. Here we present some comments regarding these four points and how they relate to Florida Scrub Jays.

Virtually all Florida Scrub Jays are nonbreeding helpers at age 1 year, and many help for several years thereafter. However, physiological maturity at age 1 year is probable because individuals of both sexes have bred at age 1 and many do so at age 2 (chapter four). The supposition that physiologically mature jays will serve as helpers is substantiated by the occasional return to helper status of a former breeder. No doubt Florida Scrub Jays delay breeding beyond the time when they are capable of producing viable gametes.

An early explanation for cooperative breeding was based on assumed delayed maturation (Skutch, 1961; Lack, 1966). This assumption apparently was based on the idea that if a population included individuals physiologically unable to breed, their best strategy for furthering their genetic lineage was to help the breeders. To our knowledge no support is given to this idea at present. We certainly do not consider that delayed maturity "explains" helping in Florida Scrub Jays, and we doubt that it actually *causes* helping in any species. Delayed maturation, where it does occur in cooperative breeders, probably evolves as a corollary to delayed breeding, as implied by Brown (1978a). Because delayed maturation represents a response to the same demographic regime as does helping, it cannot itself be viewed as a cause.

Except for brief transitory periods during dispersal, Florida Scrub Jays exist only as pairs or within family units; therefore we assume the sex ratio we obtain from counting jays in our study tract is a representative sample for the entire population. The sex ratio of breeders is equal, owing to their monogamy (chapter four). The sex ratio of yearling Florida Scrub Jays also is roughly equal. (Probably the jays produce equal numbers of each sex initially.) In contrast, the sex ratio of helpers age 2 years or older is strongly biased toward males. Clearly, sometime between age 1 year and breeding, females experience a higher mortality than males. Apparently for most bird species females disperse farther from home than males to become breeders (Greenwood, 1980). One cost of active dispersal behavior is increased mortality in comparison to living near or at home. Therefore, we suspect the preponderance of males as helpers in many species of cooperative breeders results from increased dispersal-associated mortality of females. This asymmetry in nonbreeder mortality clearly characterizes Florida Scrub Jays (see chapters seven and nine).

Some investigators postulate that certain helper systems may have arisen because sex ratios are uneven (Brown, 1974; Reyer, 1980). In all such cases, males predominate. Where breeding is monogamous, therefore, a surplus of nonbreeding males exists. Although nowhere spelled out in detail, we see this argument proceeding as follows: (1) externally imposed demographic patterns, especially differential mortality of nonbreeders, cause a sex ratio skewed toward males; (2) females become a limiting resource, so that some males (particularly younger ones) cannot breed; (3) strategies are favored by which nonbreeding males can raise their inclusive fitness, either (4a) by helping raise additional sibs (e.g., "primary helpers" of Reyer, 1980) or (4b) by gaining a competitive advantage in the contest for mates; and (5) individuals who work their way into the breeding system *by helping* gain this competitive advantage, so that helping is favored even in the absence of genetic relatedness to the nestlings.

We note that the preceding scenario suggests an extremely important, and ultimately selfish, role played by helping. The role is emphasized by Reyer (1980, p. 226) with respect to Pied Kingfisher "secondary helpers," as a means by which they "may increase *their own* reproductive success" (italics ours). The hypothesis is that helping provides a behavioral mechanism for becoming established in a neighborhood composed primarily of breeders. This establishment is so crucial that individuals will fight for the opportunity to help. Once established, access to breeding status (via obtaining space or a mate) is made easier. Thus helping serves as a means for becoming a breeder. The same role has been suggested for "standing" behavior in nonbreeding Sooty Terns (*Sterna fuscata*), who slowly work themselves into breeding colonies by competing for "standing room," and even feeding nearby chicks, within a densely packed colony (Robertson et al., 1979).

The existence of a limiting sex is critical to the above scenario. Therefore it is perhaps not surprising that colonial group-breeders have been cited most frequently as examples of this "route to sociality" (Fry, 1972a; Brown, 1974; Reyer, 1980). In such species, nesting space rarely, if ever, could limit the breeding population on its own. In contrast, our population apparently *is* space limited (chapter three). Sexually mature nonbreeders of both sexes always are present as helpers. Therefore, the skewed sex ratio cannot be construed as a *cause* of helping in the Florida Scrub Jay. Rather, we view the skewed sex ratio as a *result* of the system.

We have concluded that Florida Scrub Jay helpers do not breed. Once paired, mates rarely are more than a few meters from each other (except during incubation, after all eggs are laid), and almost never are out of signaling distance. All inhabitants of a territory are subordinate to the breeding male, and all females are subordinate to the breeding female (Woolfenden and Fitzpatrick, 1977). Many times we have seen male breeders attack male helpers attempting to dominate the female breeder. Sometimes the attacks were preceded by an alarm call given by the female. The activities of helpers around the nest are restricted to late stages in breeding, well after fertilization has occurred. Helpers do not build nests or incubate, and are vigorously driven away from the nest site by the breeders *prior* to hatching. Only under aberrant circumstances does a helper pass food to a female breeder prior to hatching. Never have we seen copulation between unpaired jays, and never have we found an egg in a nest that we suspected was from other than the resident female breeder.

As male helpers grow older, they usually increase their attentiveness to the breeding female at the nest. This attentiveness can become so pronounced that untrained observers would have difficulty distinguishing breeding males from older male helpers. (Indeed, early in the study we encountered one such example ourselves.)

Sixty-four percent of Florida Scrub Jay helpers assist both of their parents. Ninety percent help at least one parent. The replacement of lost breeders, which is by jays from outside the family group, accounts for most instances of jays helping breeders other than their parents. We assign little evolutionary significance to the exceptions where helpers help breeders less closely related than parents. However, some exceptions have been useful for teasing apart the selective forces that may drive the system. For example, situations have arisen when potential helpers had the choice of aiding nonrelatives on their natal ground or familiar close relatives nearby. Their choosing natal ground provides additional support for our opinion that occupying familiar space is of great importance to Florida Scrub Jays even before they begin to breed.

97

To summarize our thinking on the four points listed at the outset of this section, we reiterate that Florida Scrub Jays present a paradigm for the key questions regarding the evolution of cooperative breeding. Why do apparently physiologically mature individuals, regardless of their sex, refrain from breeding in favor of assisting with the breeding of others? We note that in most instances they assist their parents. However, the existence of close familial relationships between helpers and breeders within a group is not universal, and could be a by-product of the overwhelming importance of occupying familiar ground. At this point in the story, the question of how important these kinship patterns are in the system remains moot.

The Florida Scrub Jay appears to be a living example of an early stage in the evolution of cooperative breeding. Half the breeding pairs have no helpers at any one time, and helpers must acquire space beyond their natal territory before they can breed. If cooperative breeding is a derived social trait, it almost certainly came from an ancestor that bred as unassisted pairs (Brown, 1978a). A stage that probably follows the Scrub Jay's present condition would be breeding only with helpers (Verbeek, 1973, figure 19). As numerous species, including many jays, have been reported only with helpers, this stage appears to be well represented among present-day corvids. It may be an obligate behavioral trait in some species. In even more advanced stages, nonbreeders eventually breed within multipair territories, as in the Mexican Jay (Brown, 1974; Brown and Brown, 1981a) and other jays. In chapter seven we show that this system occasionally "evolves" in our own Florida Scrub Jay population.

We do not assume that the Florida Scrub Jay breeding system evolved from an ancestor that bred without helpers. Indeed, evidence exists to support an opposite conclusion.

Relatives of the genus *Aphelocoma,* and relatives of the Scrub Jay within this genus, are concentrated in the Neotropics. Cooperative breeding in New World Jays also is concentrated in the tropics. Possibly, the ancestors of modern Scrub Jays spread northward as cooperative breeders. As they spread they reduced the frequency of, and in western North America, eventually lost, the habits of cooperative breeding. The situation in Florida therefore might be a vestige of a trait that was more developed in the ancestral stock and has been lost in western North America (a possibility also noted by Brown, 1974, p. 71). This possibility does not, of course, negate the social system of Florida Scrub Jays as reflective of an early stage in cooperative breeding.

Florida Scrub Jays exhibit a rigid system of social dominance, which we suspect is of great importance in the evolution and operation of their breeding system (Woolfenden and Fitzpatrick, 1977, 1978a). It may be that the attendance of helpers to their young relatives aids in establishing and maintaining dominance over these individuals. Certainly the frequent visits to nests by helpers without food (Stallcup and Woolfenden, 1978) and the poor correlation between fledging success and food provided (Woolfenden, 1978; see chapter eight) suggest reasons other than maximizing kin productivity as the sole purpose for visiting nests. The sequence by which helpers of like sex from the same family become breeders suggests to us a mechanism by which family dominance could function in the eventual establishing of helpers as breeders (Woolfenden and Fitzpatrick, 1978a).

Dominance-subordinance systems are reported for several cooperative breeders in addition to the Florida Scrub Jay (e.g., Superb Blue Wren: Rowley, 1965; Black-backed Magpie: Carrick, 1972; Tasmanian Native Hen: Ridpath, 1972; Long-tailed Tit: Gaston, 1973; Kookaburra: Parry, 1973; Groove-billed Ani: Vehrencamp, 1978; Acorn

Woodpecker: MacRoberts and MacRoberts, 1976; Pukeko: Craig, 1976). Presumably a dominance system is among the first behavioral modifications to arise in response to the existence of potential breeders within a pair's territory (Woolfenden and Fitzpatrick, 1977), and we agree with Brown (1978a) that it may be ubiquitous among cooperative breeders.

The dominance-subordinance system in Florida Scrub Jays is rigid, but it is expressed with overt aggression only infrequently. A family of six jays, including four males, was followed for six consecutive hours as they foraged in their territory, and we witnessed only eight subtle dominance encounters (Woolfenden and Fitzpatrick, 1977). Similar results were obtained by DeGange (1976) in his thorough study of time budgets. Low frequency and low intensity of aggression have been mentioned for several other group-breeding birds (Brown, 1978a), and restraints on aggression within social breeding units were cited as a prediction of kinship theory (Brown, 1974, 1978a). However, we see no reason why similar social relationships would not evolve in *any* close-knit functioning unit of individuals. Once established, dominance-subordinance relationships need not manifest themselves in any but the most subtle behavioral interactions. Indeed, *whenever* animals must function in a close unit, restraints on overt aggression would seem to be advantageous to all the individuals.

CHAPTER SIX

Territory

SYNOPSIS

Florida Scrub Jays always reside in territories with well-defined boundaries defended year round by all group members. Territorial defense is most active immediately prior to nesting in the spring and after molt is completed in autumn. Territories typically occupy essentially the same piece of ground for many years. Ownership of territories is passed on through sequential mate replacements or through inheritance by helpers. Mean and median territory sizes equal about 9 ha (25 acres). Territory size increases significantly with family size, but not linearly. The increase becomes asymptotic among families containing two or more helpers. Some territories are consistently larger than others regardless of family size, probably owing to habitat differences. Older male helpers appear to cause significant extra growth in territory size relative to territories containing other classes of helpers. This is an important precursor to territorial budding. Minor variation in habitat quality probably affects territory sizes, but does not appear to affect reproduction in the main tract (preliminary data only). Territories disappear by becoming subsumed into neighboring ones, usually after the death of one or both breeding jays. Disappearances occur predominantly in small families occupying small territories. By helping produce extra family members helpers may insure their territory's persistence, thereby increasing their own survival and reproductive probabilities. The territory represents a land bank, the defense of which is vital to ongoing reproduction by the breeders, and to survival and later reproduction by helpers.

Florida Scrub Jays spend their entire lives dwelling in a territory. With rare exceptions, prebreeders live entirely in their natal territory. Following dispersal, jays live the remainder of their lives as breeders in one limited patch of oak scrub, which they defend the year round. The territory provides the jays' daily and annual requirements, including food, water, cover, nest sites, and nest material.

Male breeders spend about half of all daylight hours each day sitting on an exposed perch. Helpers, and breeding females except when incubating or brooding, behave similarly (DeGange, 1976). From these perches the jays detect territorial intrusions by neighboring jays, to which they respond immediately (Woolfenden and Fitzpatrick, 1977). Territorial defense is most vigorous by the breeding pair, but all family members participate. Defense of territory always ranks paramount among the daily activities. Jays often cease foraging, mobbing a predator, or attending a nest in order to confront an intruder at a territorial boundary.

Although territorial defense exists throughout the year, frequency is highly seasonal. Based on territorial call counts conducted monthly within our study tract (Barbour, 1977; figure 6.1), defense is most frequent in late winter–early spring, immediately preceding nesting, and in late summer–early fall, following the peak of molt (Bancroft and Woolfenden, 1982). The latter time coincides with a peak in dispersal frequency (see figure 7.3).

As territory is indispensable for a jay's existence and breeding, detailed information on the stability, size, and history of the land it possesses is essential for studying the social organization of Florida Scrub Jays. Appendixes B through K diagram the 224 territories we mapped over 10 breeding seasons, 1970–1979. Data taken from these maps

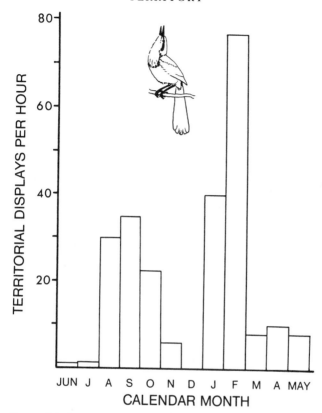

FIGURE 6.1. Seasonal distribution of territorial display calling. The August-to-November peak coincides with a peak in dispersal frequency, and the January–February peak is associated with active boundary disputes and adjustments immediately preceding the nesting season.

are used in our various analyses of how the jays partition acceptable habitat. Also mapped are the major habitats of the study tract (Appendix A) and the location of nests within each territory (Appendixes B–K).

6.1. BOUNDARY STABILITY AND PERMANENCE OF OWNERSHIP

As shown on the territory maps, extensive patches of habitat little used by the jays, especially the grassy depressions, form parts of the boundaries between neighboring territories. However, all territories have some defended borders extending through optimal oak scrub. Here, interactions between neighboring families can be observed at close range, and it is clear that such boundaries are narrow and distinct, usually less than a few meters wide.

Careful study of the territory maps demonstrates the stability of territories from year to year, as exemplified in figure 6.2. Although some shifts always occur between successive years, segments of boundaries often remain unchanged for many years. In our study, certain territories have encompassed the same ridges of prime scrub for 10 consecutive years. This stability may persist for decades. Mortality of breeders is low (see chapter nine), and ownership of the same piece of scrub therefore tends to remain with the same family lineage for long periods.

We compared successive years' nest locations to quantify the sedentary nature of Scrub Jay breeders. We chose this measure because nests always are the focal point of a family's activities during a few months of each year. For each family we measured distances between the first nest ever

FIGURE 6.2. The boundaries of a territory (SAND) that was occupied by one male and three successive mates for eight consecutive years. Budding (see FIRE territory in Appendix C) established the territory in 1972. The male died after breeding in 1979. In 1978 territorial expansion (heavy dashes) south of a big pond resulted in a son budding off a territory (SSAN) for the 1979 breeding season. For reference a sand road (small dashes) and several temporary ponds (stippled areas) are shown. (See Appendixes D–K for further details.)

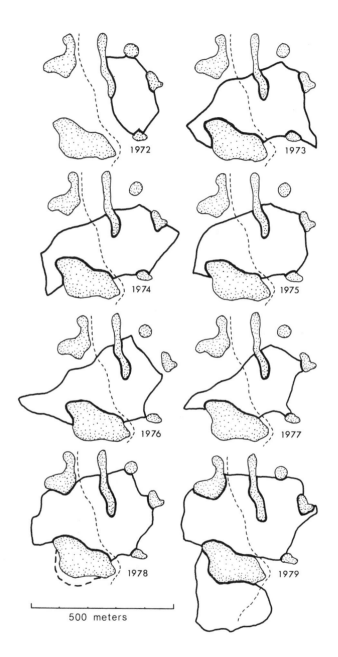

1972

1973

1974

1975

1976

1977

1978

1979

500 meters

found for a given territory and first nest sites of each subsequent breeding season. As a control, we measured distances between successive nests of the same family during
one breeding season, where the pairs and their territories
remained unchanged. As the years passed, a chronological
sequence of replacement mates occurred in most territories, especially those measured over the longest time span.
In part because of this, as shown in figure 6.3, some drifting
away from the original nest sites occurs with time, so that
after nine years the average nest is placed twice as far from
the original as it is in the second year. The fluidity of territorial expansion and subsequent contraction also contributes to this drifting. Nevertheless, the mean distance between
nest sites even eight years apart (288 m) is scarcely more
than twice the mean distance between successive nests of
the same season (138 ± 80 m). This amount of drifting is
substantially less than the diameter of an average-sized territory (9.0 ha, diameter = 339 m where circular). This
result graphically demonstrates the sedentary habits of
Florida Scrub Jay pairs.

6.2. THE AVERAGE TERRITORY

Our calculations of territory size exclude major areas of
totally unused habitat, especially the grassy depressions, but
include some expanses of mixed palmetto and oak. The
latter habitats often grade gradually onto the ridge tops
covered by oak scrub, the essential habitat requirement for
all territories (chapter three). As shown in figure 6.4 the
area of oak scrub contained within a territory varies directly, albeit imperfectly, with territory size. In the following analyses we assume that our territory size measurements
accurately reflect the relative amount of usable land held
by each family.

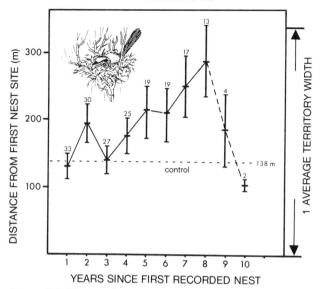

FIGURE 6.3. Distances between first known nests within pairs' (including their staggered replacements) territories, and all their subsequent seasonal first nests. Average distance between first and subsequent nests in the same territory during one nesting season is shown as a control. Although some drifting away from the original site occurs, even after 8–10 years nests are placed no farther from their initial site than the mean diameter of an average territory. (Numbers above the standard error bars show the sample sizes.)

Table 6.1 (see p. 109) summarizes the sizes of 221 breeding-season territories over nine years. A frequency distribution of the overall sample is shown in figure 6.5. Territory size has ranged from 1.2 to 20.6 ha, with a mean of 9.03 and a median of 8.68 ha. The size distribution does not differ significantly from normality ($p > .05$; Kolmogorov-Smirnoff test).

Mean territory size is similar for all years except one. The significantly smaller size in 1972 occurred because an unusually large number of new territories was established that spring, and new territories always are small. As discussed elsewhere, this occurred two years after the most

107

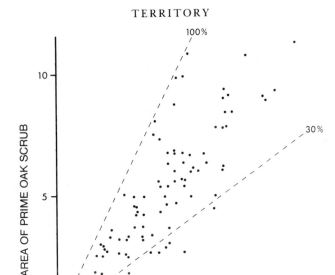

FIGURE 6.4. Relationship between territory sizes and the area of oak scrub contained within those territories. Data points (N = 90) are from three breeding seasons, 1977–1979. Oak scrub is the essential habitat for the jays, and total territory size is a good, though imperfect, predictor of the amount of this habitat contained therein (r^2 = .67). Maximum possible (100%) and minimal (30%) proportion lines are shown for reference.

successful breeding season thus far encountered during the study.

6.3. TERRITORY SIZE AND FAMILY SIZE

With smaller sample sizes, we showed previously (Woolfenden and Fitzpatrick, 1978a) that territory size and family size are positively correlated, apparently through a causal relationship: as families grow, so do their territories. With

TABLE 6.1. Florida Scrub Jay Territory Sizes in Hectares, 1971–1979

	Breeding Season									Total	Annual Mean
	'71	'72	'73	'74	'75	'76	'77	'78	'79		
N	14	22	24	19	26	25	27	31	33	221	9 yr.
Mean	7.9	6.8[b]	8.0	9.1	9.6	10.4	10.3	9.5	8.4	9.0	9.0
S.D.	3.7	3.6	4.1	3.9	4.1	4.3	4.0	4.2	4.1	4.1	1.2
Range	2.9–15.3	2.6–19.8	2.7–19.2	4.6–18.7	1.2–17.7	2.6–18.4	3.2–17.6	2.3–18.8	2.1–20.6	1.2–20.6	2.7–18.5
Density[a]	3.73	4.04	3.88	3.59	3.79	3.72	3.61	3.83	4.00	—	3.81

[a] Density is measured in pairs per 40 ha, based on core area of study tract (table 3.2).
[b] 1972 mean territory size significantly smaller than pooled sample of all other years (Student's t, $p < .01$); no other year so differs (see figure 3.18).

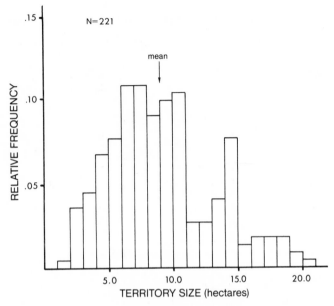

FIGURE 6.5. Frequency distribution by size of 221 breeding-season jay territories. The distribution does not differ significantly from normality. Mean size is 9.0 hectares.

the data now available, we can examine the relationship between jay families and their territories in greater detail. Table 6.2 presents territory sizes for Florida Scrub Jay families that varied in size from 2 to 8 individuals. Each family's territory size is plotted separately against family size, for each year it was measured, in figure 6.6. These data indicate a trend toward larger territories among larger families. However, several complications exist in these data, not the least of which is the extremely high variance in territory sizes within any one class of family sizes. Therefore, we here examine the data several additional ways in order to evaluate more carefully the family size–territory relationship.

TABLE 6.2. Territory Size as Related to Family Size in the Florida Scrub Jay

Family Size	N	Territory Size (ha)		
		Mean	S.D.	Range
2	92	7.2	3.2	1.2–14.6
3	62	9.8	4.4	2.6–19.2
4	42	10.2	4.1	3.2–19.8
5	16	11.6	4.3	4.1–20.6
6–8	12	11.6	2.5	7.5–14.5
3–8	131	10.4	4.3	2.6–20.6
Bud	16	4.5	2.9	2.1–6.8 (14.1)[a]
Total	224	9.0	4.1	1.2–20.6

[a] Range shown for N = 15; size of the one exceptionally large territorial bud shown in parentheses.

Relative Family Size

While the regression line in figure 6.6 is positive and highly significant ($y = 1.2x + 5.2$; $F = 30.8$, $p<.001$), it does not account for much of the variance in the scatter plot ($r^2 = .12$). One complication is that while breeder density and average territory sizes remain relatively stable from year to year (figures 6.2, 6.3), average family sizes do not. For example, in certain lean years a family of 3 may be able to expand its territory at the expense of less successful pairs, while in other, more populous years 3 birds per group will be small with respect to the neighboring population. Therefore, a measure of *relative* family size should improve upon the absolute sizes used in figure 6.6. We plotted territory sizes against several such relative measures for family size. The scatter about the regressions remained virtually unchanged, indicating that factors besides simple family size are helping to determine territory size.

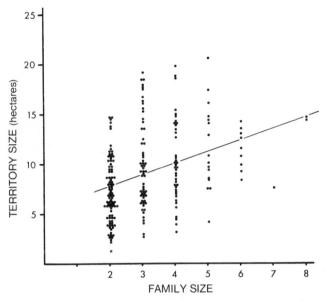

FIGURE 6.6. Relationship between family size and territory size for 221 breeding-season jay territories. Linear regression is shown (r^2 = .12).

Furthermore, our using the regression to test for an increase implies that we expect such an increase to be linear. As shown by table 6.2, and below, the assumption does not appear to be true.

Nonlinear Increase in Territory Size

Figure 6.7 graphs mean territory sizes and their standard errors for the data summarized in table 6.2. We suspect that the asymptotic rise in territory size with family size is real, although *t*-tests between successive classes show a statistically significant increase only between family sizes 2 and 3 ($p < .001$). Again, high variances in each sample indicate that additional factors complicate the relationship implied by figure 6.6.

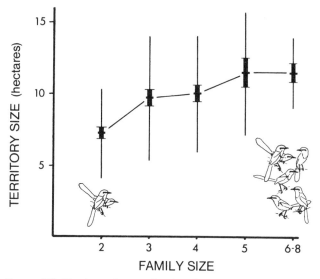

FIGURE 6.7. Territory size as related to family size based on data summarized in table 6.2. Shown are the means (thick horizontal lines), standard errors (vertical bars), and standard deviations (vertical lines).

The Cause of Territorial Growth

A positive relation between territory size and family size is insufficient evidence for a causal relationship. Larger territories could harbor individuals or resources capable of producing more young than in smaller territories. As discussed in Woolfenden and Fitzpatrick (1978a), the critical tests require measurements of territory size *within* families that have varied in size over several years. We now have 30 families that meet these criteria, and we plot their territorial histories two ways in figure 6.8a, b. In figure 6.8a, a line connects territory size values associated with maximum and minimum family size values for each family. In 27 of 30 cases, territory size was larger when the family was larger. Tested against the zero-growth null hypothesis

113

that as many should show decreases as increases, this result demonstrates a highly significant intrafamilial increase in territory size as family size grows ($\chi^2 = 19.2$, $p < .001$). The same test can be applied to the data plotted in figure 6.8b, where territory size values at each increment of familial growth are plotted and lines connect values for individual families. Of 25 families, 22 (88%) show larger territories at family size 3 than at 2. Of 23 families, 17 (74%) show larger territories at family size 4 than at 3. These values represent significant growth in territory size between 2 and 4 individuals ($p < .001$, $p < .05$, respectively). At family sizes greater than 3, the same pattern of growth is evident, but sample sizes are small and the results do not attain statistical significance. Average percentage increases between each family size from 2 to 5–8 are shown separately in figure 6.9. Again, the results show that the most important growth in territory size occurs between family sizes 2 and 3. Growth continues, more or less asymptotically, thereafter (cf. figures 6.7 and 6.9).

Interfamilial Variation in Territory Size

Figure 6.8 reveals the most important source of variation in the territory size–family size relationship, namely that variation is much smaller *within* families than *between* families. The series of lines in figure 6.8a show roughly parallel increases in territory size with family size *even though some families consistently occupy relatively much larger territories than others*. We suspect that these consistent interfamilial differences relate to subtle variations in the quality of oak scrub

FIGURE 6.8. Changes in territory size associated with minimum and maximum family sizes (6.8a), and with each increment in family size (6.8b) for 30 jay families. Where several years' data were available for a family of a given size, the mean of those territory sizes is shown.

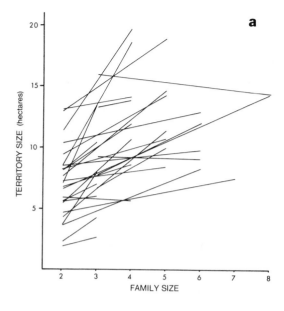

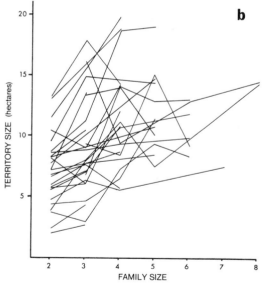

owned by different families, a conjecture that remains untested. In any event, the pattern that emerges from figure 6.8 is that territory size increases with family size, but only within the confines of small variations as compared to total variation within the population as a whole. One corollary of this pattern is a highly significant relationship between territory size in a given year and its size the previous year (figure 6.10). Future work on territoriality will concentrate on the question of why some territories remain inherently larger than others regardless of their family histories.

Stability of Territory Sizes in Stable Families

The possibility remains that territories simply have increased in size from one year to the next and that the results presented above are caused by this feature combined with coincidental additions of family members. However, in a stable population where breeding density remains unchanged over a decade, the number of territories that de-

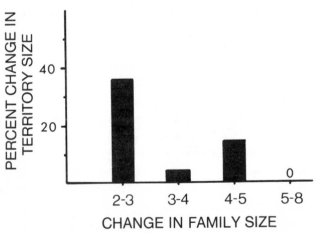

FIGURE 6.9. Average percentage increases in territory size with change in family size.

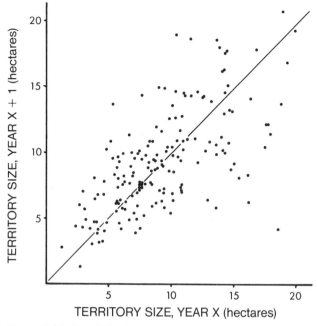

FIGURE 6.10. Correlation between territory sizes one year and the next year for the same jay family. First principle axis is shown through the scatter plot.

crease in size should roughly equal the number that increase. The actual data show 82 increases, 77 decreases, and 19 territories that remained within .4 ha (1 acre) of the same size from one year to the next. In a more refined test, we compared changes in territory size between consecutive seasons within families whose size and breeding pair remained unchanged. Of 43 such cases, 19 showed slight territory size increases, 17 showed slight decreases, and 7 did not change appreciably. The mean size change for the entire sample was $-.4 \pm 2.6$ ha. These data support the null hypothesis that no consistent increase or decrease in territory size occurs without a change in family size. Territorial growth occurs at the expense of neighboring territories.

Significance of Territorial Growth

The process of territorial budding (Woolfenden and Fitzpatrick, 1978a), whereby a helper male with a female from outside his family become sole occupants and breeders in a segment of the territory in which he helped, is described in chapter seven. Territorial growth occurs between the first and second breeding seasons after budding. Nine territories mapped during both their first and second years of existence averaged 3.6 ± 1.5 ha the first year and 6.2 ± 2.2 ha the second. All nine territories were substantially larger during their second year of existence (percentage increases ranged from 28 to 124, $\bar{x} = 179$ percent; trend significant, $p<.01$, Wilcoxon test). During their second year of existence these nine territories were not significantly different in size from territories of all other unassisted pairs.

The pattern developed thus far is that changes in territory size parallel changes in family size. As families grow, their territory grows; as family members disperse or die without replacement, territory size shrinks as neighboring, more successful families expand. Territories of families that remain stable in size have an equal probability of growing or shrinking, depending upon the events in neighboring territories. As we show below, however, one additional variable—the sex of helpers within the family—also affects territorial growth patterns. Because of an important sexual difference in dispersal strategies, namely, the possibility of territorial budding by male helpers, this additional complexity is extremely important to the broad question of what helpers may gain by remaining home.

118

6.4. SEX OF HELPERS AND TERRITORIAL GROWTH

A powerful test for whether helpers of different sexes have different effects on territorial growth patterns would be to measure territory sizes of a given pair during years when helpers represent only one sex or the other. Because territorial growth is gradual, several successive years would be required with helpers of one sex, several years with the other sex, and several "control" years when no helpers were present. Unfortunately, our population study has not continued long enough for this type of longitudinal comparison, given the complexities introduced by mate replacement, dispersal, and mixed-sex clutches that characterize the social system. Therefore, at present, we can make only cross-sectional tests in which we compare average sizes of families with helpers of different constituency. The results suggest that when the more desirable longitudinal analyses can be made, they will support the hypothesis that male helpers cause more territorial growth than do female helpers.

Table 6.3 makes four territory-size comparisons among families with different helper constituencies. (1) Among families that contain only one-year-old helpers, no difference in mean territory size exists whether the helpers are male, female, or mixed, although all three classes show significantly larger territories than do simple pairs. Hence, if a sexual difference exists, it is expressed only as helpers become older, a result that would parallel the findings of Stallcup and Woolfenden (1978) for feeding rates of helpers. (2) Among families with at least one helper, those that include an older male helper (age 2 or more years) have larger territories than those lacking an older male. The difference, however (11.3 vs 9.8 ha), falls just below statistical significance (Student's t, $p = .07$). (3) To distinguish more effectively the relative effects of older male from

119

TABLE 6.3. Territory Size as Related to Sex and Age of Helpers

Family Size	Helper Constituency	Territory Size (ha)		
		Mean	S.D.	N
≥3 ($\bar{x}$ = 3.3)	yearling ♂♂ only	9.7	± 4.2	23
≥4 ($\bar{x}$ = 4.9)	yearling ♂, ♀ mixed	10.4	± 4.0	13
≥3 ($\bar{x}$ = 3.3)	yearling ♀♀ only	10.4	± 4.4	20
≥3 ($\bar{x}$ = 4.2)	older ♂ present	11.3[a]	± 4.3	40
≥3 ($\bar{x}$ = 3.8)	no older ♂ present	9.8	± 4.1	91
≥3 ($\bar{x}$ = 4.0)	incl. older ♂, no older ♀	11.5[b]	± 4.0	33
≥3 ($\bar{x}$ = 3.9)	incl. older ♀, no older ♂	9.2	± 3.2	22
3	older ♂	11.6[b]	± 4.7	14
3	older ♀	7.9	± 3.3	11

[a] Difference in territory sizes with and without older male helpers nearly significant ($p \approx .05$).
[b] Significantly larger than corresponding value for females (Student's t, $p < .05$).

older female helpers, we compare families that include older males *but not* older females with those lacking older males *and containing* older females. Territories are significantly larger in the sample containing older males (11.5 vs 9.2 ha; $p < .05$). It is necessary to point out that overall family sizes within each of these two samples are equal, ensuring that the larger territories are not caused by larger groups that happen to contain older males more frequently. (4) Finally, we further refine our testing by comparing families of three that include only one, older helper. Those in which the lone helper is an older male (N = 14) have significantly larger territories ($p < .05$) than those in which the helper is an older female (N = 11).

The trend toward larger territories among families with older helper males is perhaps not surprising, as males tend to be more active in territory defense than females, even at the helper stage (Woolfenden and Fitzpatrick, 1977). However, as shown in chapter seven, territorial budding by male helpers occurs in large territories, which presumably were enlarged in part through the territorial efforts of the helper himself. This observation strongly suggests that many male helpers actually reap some direct reproductive benefits from the energetic investments they contribute as helpers.

6.5. VARIATION IN TERRITORY QUALITY

In chapter three we showed that tall, unburned areas of scrubby flatwoods bordering the main study tract were progressively abandoned by Florida Scrub Jays during the 1970s. Prior to their disappearance, the jays experienced lower reproductive success in this poor quality habitat compared to jays in the main study tract (table 3.1, figure 3.15). This clearly demonstrates that appreciable variability in habitat quality can exist between occupied territories. It is logical to suspect that some such variation exists on a smaller scale within the main tract. In the absence of detailed measures of this variation (a field project now under way) we can offer only subjective impressions on the topic at this point. As measured by reproductive success, the variation in territory quality does not appear to be great.

Our most important statement about habitat variation is that patches of oak scrub appear to have insufficient variability to cause jay families to compete actively for certain, obviously superior areas. In certain other cooperative breeders (e.g., several babblers: Brown and Balda, 1977; Pukeko: Craig, 1980a,b; Acorn Woodpecker: MacRoberts and MacRoberts, 1976; Stacey, 1979a,b; Koenig and Pitelka, 1981) groups show evidence of active competition for clearly favorable habitat patches or other limited resources such as mast-storage areas. However, Florida Scrub Jay groups occupy essentially the same patch of scrub for many years in succession (e.g., figure 6.3), while territory boundaries merely fluctuate as neighboring family sizes change. Major territorial shifts do occur, especially when families break up (see section 6.6. "Disappearance of Territories"), but the shifts do not show obvious relation to any habitat feature beyond the requirement that oaks be present in any newly acquired territory segment.

We suspect that most functional habitat variation among territories falls into two categories:

1. A scan of the habitat and territory maps in Appendixes A through K shows that territories contain differing proportions of the various open microhabitats, including little-used ones. (Figure 6.4 showed the resulting imperfect correlation between territory size and area of oak scrub contained therein.) As described in chapter three (3.2. "The Scrub Habitat"), some areas in the study tract contain large grassy depressions, palmetto flats, or a mixture of palmetto and low, widely spaced oaks. The average size of territories that encompass these areas is larger than those situated within broad, continuous stands of oak.

2. Although all territories contain considerable stands of pure oak scrub, the height and apparent richness of these stands vary slightly. The patchy nature of scrub fires combined with slight elevation and soil differences between ridge tops undoubtedly account for much of this minor variation. Probably there exists a postfire successional stage at which the oak scrub is optimal for Florida Scrub Jay reproduction and survival. As discussed above, when a rich stand of oaks is left unburned too long, it becomes unusable. Thus, through time the scrub habitat represents an ever-changing mosaic of slightly differing oak patches, whose

exact configuration is influenced by the fickle fluidity of Florida forest fires.

With only ten years' results, we cannot yet test the effects of these minor habitat differences on survival and reproduction while controlling for individual variations between pairs, effects of helpers, and other factors. A preliminary analysis (figure 6.11) shows remarkable uniformity in average fledgling production throughout the study tract. We suspect that the most significant impact of habitat variation is its relation to territory size differences, but this too remains untested.

6.6. DISAPPEARANCE OF TERRITORIES

Most territories persist for many years, and probably for decades, even though ownership of them changes hands through deaths, mate replacements, and inheritance. New territories periodically are created primarily through territorial budding (see chapter seven). By the same token, territories can disappear. Whereas territorial budding occurs as a family builds in numbers and landholdings, the disappearance of a territory occurs most often when a pair or small family occupies a small territory. Almost universally, the proximate cause of territorial disappearance is death of at least one member of the pair.

Table 6.4 summarizes pertinent statistics regarding the 23 instances in which well-defined territories have disappeared between successive breeding seasons. Sixteen of these cases (70%) involved pairs without helpers, and in no instance was more than one helper present. All helpers were yearlings, and five of the seven were females. In cases where both members of the pair died before either could pair again, their territory sizes had not differed from the population's average. However, none of the remaining 14 ter-

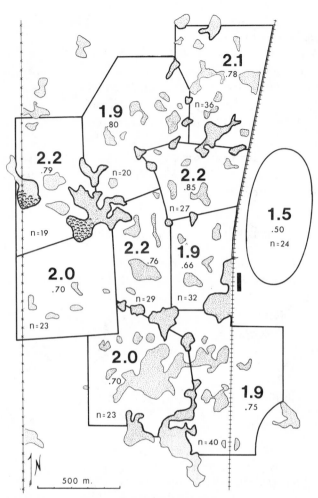

Figure 6.11. Fledgling production (large numbers) and nesting success (decimals) for all families studied during the 1970s in nine randomly drawn geographic segments of the study tract and in the tall, dense unburned scrubby flatwoods "on the other side of the tracks" (oval figure). Excluding the latter, artificial habitat, reproduction is remarkably uniform throughout the tract.

TABLE 6.4. Events and Statistics Pertaining to Territorial
Disappearances

Events	N	Mean Territory Size (ha)	Yearlings That Moved[a]
♂ and ♀ die	9	9.6[b]	3 ♀♀, 1 ?
♂ dies, ♀ replaces elsewhere	8	4.8	2 ♀♀, 1 ♂
♀ dies, ♂ replaces elsewhere	3	4.0	—
♂ and ♀ split, return home	2	2.9	—
♀ divorces, ♂ returns home	1	2.7	—

[a] No territory contained more than one helper; all were yearlings.
[b] Mean territory size where both breeding individuals die approximates
overall mean for population; mean size of all others combined (4.2 ha)
significantly smaller than overall population mean (Student's t, $p < .01$).

ritories was larger than 7.0 ha, and their average size (4.2
ha) fell significantly below the overall mean.

Territories physically disappear because neighboring
families subsume their land holdings (see Appendixes B
through K). Sometimes this occurs when the widowed jay
pairs with a widowed or budding jay from an adjacent
territory, but this is by no means the rule. Lone, widowed
jays or solitary helpers, especially females, may wander and
join established groups far from their lost territory. Territorial disappearance is rare when lone males (breeders or
helpers) persist after a death in the family. This observation
is consistent with the general pattern that successful site
fidelity, including land inheritance, is highly skewed toward
males. However, for all individuals, regardless of age or
sex, many advantages exist in retaining the same, familiar
piece of land after a breakup in family structure. This truism
is significant, given that territorial disappearance involves
only small families, usually occupying the smallest territories. We propose that helping to raise additional family
members provides some measure of insurance that the ter-

ritory will persist in the event that one or both breeding individuals should die.

6.7. ADVANTAGES OF LARGE TERRITORIES

Florida Scrub Jay territories are large for the size of the bird. As reported by Hardy (1961) and Atwood (1980a,b), two different subspecies of Scrub Jays in western North America live in territories averaging one-fourth the size of those in our study population. In comparison with a variety of omnivorous birds' territories (Schoener, 1968), the Florida Scrub Jay shows relatively the largest territory for its body size.

Food Supply and Minimum Territory Size

The all-purpose territory has long been viewed as one means by which individuals can ensure themselves of an adequate supply of some resource (Howard, 1920). Often the key resource is food, and many studies since the pioneering one by Stenger (1958) show that territory size is at least roughly adjusted according to the density of food within the local area (reviewed by Davies, 1978, and Myers et al., 1979). However, various lines of evidence lead us to suspect that Florida Scrub Jays habitually live in territories substantially *larger* than would be required if food were the sole determinant of their size.

1. Many pairs of jays have lived for several consecutive years in territories significantly smaller than the average size for a simple pair. The best example involves a male who had a slightly, but permanently, crippled wing. In three years (1973, 1975, 1976, Appendixes E, G, H, STIL) his territory measured 2.7, 1.2, and 2.6 ha (the territory disappeared in 1974 following a divorce). The territory

contained a pair in 1973 and 1975, and a pair plus a helper in 1976. The pair fledged two young in 1975. In this and numerous other examples, no clear habitat difference could explain the unusually small territories. Usually, the sizes resulted from social circumstances (e.g., newly formed pair, injured breeder, large family sizes in surrounding territories, etc.). The point of such examples is that in terms of food alone, a pair, and even a family, probably does not require between 7 and 18 ha of usable land in order to survive during most years.

New territories almost always begin small ($\bar{x} = 4.2$ ha). In most cases, they grow to nearly average size within one to two years after forming (see chapter seven). However, in four instances new territories were abandoned after a single nesting season (twice resulting in abandoned nest attempts), despite both breeders surviving. These four "territories" varied in size from 2.1 to 2.9 ha ($\bar{x} = 2.5$ ha), and probably represent cases with insufficient potential for stability and growth. Six other, similarly sized territories survived and grew. This size range thus represents our best guess for the minimum size necessary for a viable territory.

2. Within a given territory, between-year variation in territory size is considerable, even when the variation caused by changes in family size is removed. These increases and decreases from year to year are not correlated among territories, suggesting that annual variation in food abundance (over the habitat as a whole) does not cause their occurrence. Again, the social circumstances are strongly implicated, and individual case histories bear this out.

3. No clear formula exists whereby we can calculate how much area is required *per individual jay*. This is because, as additional family members are added, territory size increases asymptotically (figure 6.7). If land area were fine-tuned to the food needs of individuals, we would expect

more uniform, perhaps linear, adjustment of territory size to family size.

Possibly the observation that average territory size reaches an asymptote at 5 or more individuals, combined with the rarity of families greater than this size, suggests that food limitation might become important at this family size. If so, it would imply that each jay requires a minimum of just over two hectares (11.6 ha/5 to 6 individuals). Therefore, a pair of jays would require about 4.5 hectares. Perhaps not coincidentally, this matches the average size of newly formed territories (see chapter seven, table 7.2). The fact that simple pairs live in territories averaging nearly twice this size certainly suggests that food alone does not play the critical, ultimate role in determining territory size.

Territory Size and Fledgling Production

A slight positive correlation exists between territory size and number of fledglings produced (linear correlation, $r = .18$; 218 d.f., $p < .01$). However, this relationship appears to arise because the larger territories tend to be held by families with helpers, and the presence of helpers increases fledgling production (see chapter eight). When variance in family size is removed through partial correlation analysis, the relationship between territory size and fledglings produced disappears ($r = .09$; $p > .05$). Although this result must be interpreted cautiously, it does suggest that territory size does not itself affect fledgling production very much. Indeed, table 6.5 shows that no significant correlation exists once pairs are segregated into samples with versus without helpers. Therefore, whereas larger families do produce more fledglings, it does not appear that this results from these families owning more land or more resources. The large territories of Florida Scrub Jays have not arisen through their effects on offspring production.

ᴛᴀʙʟᴇ 6.5. Fledgling Production in Relation to Territory Size,
ʜ and without Helpers

	Territory Size (ha)					Correlation Coefficient
	0–4.0	4.1–8.0	8.1–12.0	12.1–16.0	16.1–21.0	
ʜ helpers						
ᴇdglings per pair	2.0	2.3	2.3	2.8	2.2	0.04[a]
).	1.4	1.4	1.5	1.6	1.7	
	4	42	41	29	15	
ᴡhout helpers						
ᴇdglings per pair	0.8	1.6	1.4	2.7	—	0.19[b]
).	1.1	1.5	1.3	0.8	←	
	15	38	30	6	—	

ᴄorrelation between territory size and fledgling production not significant; ᴛatistic = .17, d.f. = 1, 129, $p < .20$.
ᴏrrelation not significant; F-statistic = 3.27, d.f. = 1, 87, $.05 < p < .10$.

Extra Birds

One explanation for unusually large territories is that more than two jays regularly occur within them. As shown in a previous section, group size and territory size are positively correlated. Large territories proximately arise through the relatively greater territorial activity of larger families. When helpers disperse, pairs or small families often are left with enlarged territories for some time thereafter. This raises the average size (and increases the variance) for territories held by smaller families.

Territory as a Land Bank

Large natal territories are advantageous to helpers in several ways related to the quest for breeding space. As we show in the next chapter, males are more likely to bud a new territory of their own from larger territories. For this reason, gaining and holding extra land in the form of a larger territory can be viewed as an investment by male helpers to increase their own future breeding potential. Such an advantage would take the form of some measurable increase in the probability of gaining familiar breeding space that results from territorial growth caused by the presence of extra birds. Viewed in this fashion, the large Florida Scrub Jay territories emerge as a form of "land bank" invested in, and often inherited, by the helpers (Woolfenden and Fitzpatrick, 1978a).

It must be noted that this kind of advantage need not accrue to every individual in the real world for it to represent a positive selective influence. We suggest merely that the potential for inheriance always exists as a *probability*, and that male helpers enjoy a net increase in this probability as a result of any territory size increase. In view of the positive influence that helpers have on breeders' reproduction (see chapter eight), this "land bank" hypothesis clearly implies the existence of a direct reward for helping (contra Brown, 1978b).

Numbers of Neighbors

In a densely packed community of irregularly shaped territories, larger territories are likely to be bounded by more neighbors than smaller ones. Data supporting this conjecture are presented in figure 6.12. (The figure also shows that territories are packed approximately hexagonally.) We can envision one important advantage to having a greater number of neighbors abutting the territory. Again the advantage relates to dispersal, and this advantage applies equally to male and female helpers.

Dispersal forays outside the natal territory are dangerous. The higher mortality rate of females (see chapter nine) attests to the greater hazards facing individuals that must leave their territory to locate a breeding vacancy. It follows that any mechanism reducing a dispersing individual's de-

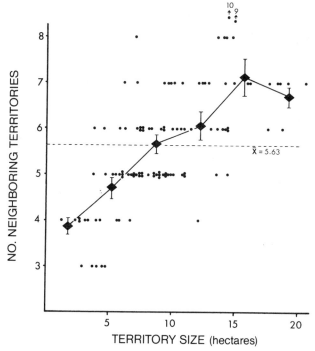

FIGURE 6.12. Relationship between territory size and number of contiguous neighboring territories. Means and standard errors are calculated at 4-ha (10-acre) intervals starting with 2 ha (5 acres).

pendence upon extraterritorial forays will confer a survival advantage to that individual and raise its probability of dispersing successfully. The data shown in figure 6.12 suggest that helpers dispersing from larger territories should be more likely to encounter a breeding vacancy without leaving home, simply because they can monitor more neighbors from within the territory. Table 6.6 bears this out. Significantly fewer helpers disperse farther than to an adjacent territory when they originate from larger territories.

131

TABLE 6.6. Territory Size and Short-Distance Dispersal in 48 Florida Scrub Jays Who Did Not Inherit Breeding Space

	Territory Size (ha)					
	3.5–7.0	7.0–10.5	10.5–14	>14	<9	>9
Dispersers (N)	9	16	8	15	15	33
Prop. adjacent to home	.22	.44	.50	.53	.27*	.52*

NOTE: Sample includes 15 males, 33 females.
*Difference between these proportions significant ($p < .05$).

Conclusive data proving the advantage of nearby dispersal over more distant attempts would consist of survival probabilities both before and after dispersal attempts (successful and unsuccessful). Necessary sample sizes are not yet available for this stringent test. However, we have shown that larger territories increase the probability of short-distance dispersal, apparently because more adjacent territories are available to the disperser. We speculate that this confers a real advantage upon individuals in larger territories. Increased probability of short-distance dispersal adds another benefit to helpers who can increase their territory size, hence their own potential landholdings and dispersal probabilities, by helping to increase the size of the family with which they reside.

6.8. CONCLUSIONS

Why are Florida Scrub Jays *so* territorial? The question recurs with every walk through our study tract. For visitors during any month of the year, we often draw a line in the sand at an appropriate spot, and by attracting neighboring jay families we promote an all-out terrtorial battle. Members of both families use calls, postures, ritualized flight displays, and occasional brief fights to advertise their ownership of land on either side of an imaginary line as thin as our boot mark in the sand. Almost any activity, including nest tending, will be temporarily abandoned by jays to display against intruding neighbors. Brown (1963) remarked that in his experience Scrub Jays in southwestern United States exhibited more active territoriality than any other jay. By all accounts and our own experience, the Florida population may exceed even the western Scrub Jay in the vigor of its territoriality.

Two major features of their biology jointly favor strong territoriality in Florida Scrub Jays. The first pertains to habitat structure and the jays' typical behavior within it, a relationship that bears all classic features of relatively easy resource defendability (Brown, 1964; Brown and Orians, 1970). The second point is that land itself represents a critically limiting resource, well worth the energy required for its defense (chapter three).

Prime habitat for Florida Scrub Jays consists of low vegetation with only a few scattered trees. Furthermore, in our study tract the oak scrub is interspersed with little-used grassy depressions whose borders are used by the jays to form buffered boundary segments that require less vigorous defense. Although the jays typically forage in concealment on or near the ground, they also spend over half their waking hours perched on high shrubs or trees that afford a full view of their nearly two-dimensional surroundings. From these perches, jays easily spot intruders anywhere along the territory boundary and can fly unobstructed to the scene. The open understory and bare sand substrate of the scrub afford little chance for a detected intruder to escape from view. In short, the oak scrub is physically easy to defend, especially by groups of medium-sized, intelligent corvids that spend much of the day perched in full view of nearly every piece of their 9 ha home range.

Even more important than habitat defendability is the overwhelming importance of owning land. It is necessary for breeding, yet every square meter of usable scrub is constantly filled and more jays always exist than are physically permitted to breed at any one time. "Privately owned" scrub of sufficient area provides food, shelter, and nesting resources, and it also confers breeding privileges upon the landed pair. With a surplus of jays always competing for breeding space, in a habitat whose physical defense is relatively easy, the benefit-cost ratio of strong territorial defense clearly favors its existence among breeding Florida Scrub Jays (Davies, 1978).

Less clear to us, and to students of social behavior generally, are (1) the selective pressures leading resident, territorial breeders to permit extra birds to reside peacefully in the same territory, and (2) the reasons behind Florida Scrub Jays' defense of such relatively oversized territories. We believe these two questions to be intimately related. A simple explanation for larger territories, which was suggested first by Norris (1958), is that more than two individuals frequently do live in them, necessitating defense of more space than a simple pair would require. The positive relationship between territory size and group size certainly supports this idea; however, we suspect that this is only one part of the story.

We view land as a resource that can be sequestered even for unborn young. In a population where offspring remain at home, especially if they cause territories to grow, even the simple pairs are forced to defend more land than they require on their own simply to live and reproduce. A key feature of the territorial system lies in the potential for some birds to inherit portions of their familial land. The existence of inheritance results in an endless competition between lineages, generation after generation, simply for breeding space. A pair without helpers, defending only enough land to survive and reproduce, cannot win in competition with pairs that defend enough extra land to provide a core for future generations' territories, particularly when some pairs are able to expand their territories through the additional efforts of helpers. In such a system, parents provide for their offspring, including ones yet unborn, by defending more area than they themselves need. Offspring improve their own fitness by adding to the land holdings, which increases both their opportunity to inherit space and the ease with which they can safely disperse to a neighboring territory to fill a vacancy.

It does not escape our attention that the model just presented introduces a paradox. On the one hand, we note that the breeding community is *saturated*, and that more birds are present than can be permitted to breed at any one time. On the other hand, we propose that Florida Scrub Jays typically defend more land than they require for survival, suggesting a permanent condition *well below saturation*, and implying that the territorial strategy itself controls breeding population density. We escape the paradox, in the short run at least, by clarifying our definition of "saturation." The concept of saturation implies an equilibrium condition, where a constant, intrinsic tendency to increase in numbers cannot be expressed because of some limiting resource. To Florida Scrub Jays the limiting resource is land. The mechanism enforcing this limitation is territoriality. The exact size of territories determines the saturation density. Long-term reproductive strategy in the presence of habitat limitation determines the sizes of territories. What remains is to explore in detail the implications of total habitat limitation (i.e., enforced zero-population growth) upon these long-term reproductive strategies. Although we have made some headway in this chapter, more spirited analyses of this important and little-studied theoretical issue are postponed to chapter ten.

CHAPTER SEVEN

Dispersal

SYNOPSIS

Florida Scrub Jay helpers obtain breeding space by any of four ways: (1) through mate replacement in territories outside their natal one; (2) by "territorial budding," in which a male, paired with an immigrant female, inherits a portion of his natal territory; (3) through direct inheritance of the natal territory following breeder deaths; and rarely, (4) by establishing a territory de novo *between existing ones. About 56 percent of breeding males obtain breeding space through some form of territorial inheritance. About 94 percent of breeding females pair outside their natal territories. Helpers rarely disperse in sib groups. Females disperse substantially farther than males, but neither sex tends to disperse more than a few territories from the birthplace. Dispersal movements show two peaks annually, one associated with the spring breeding season, the other following molt in autumn. Long-distance dispersal is limited to females, apparently peaking in the autumn of their second year. Females temporarily leave their natal territories to monitor surrounding territories in search of available males. Juveniles show tendencies to wander temporarily from their natal territories during the spring and fall dispersal peaks. Territorial budding follows territorial expansion, with the breeding pair frequently participating strongly during the expansion phase. Budding represents the principal means by which new territories appear, and is virtually restricted to males 2–4 years old ($\bar{x} = 3.1$). Multipair territories are rare. They represent an extension of the budding* *system and suggest an evolutionary route to the more advanced communal breeding systems of tropical New World jays. Effective population size is small in comparison with most bird populations, but inbreeding coefficients are not sufficiently high to account for the evolution of cooperative behavior in this species. Cooperative breeding apparently evolved in response to limited opportunities for successful early dispersal from the home territory.*

A major event in the life of any organism is its transition from nonbreeding to breeding status. This change is especially interesting in the Florida Scrub Jay because, as shown in previous chapters, young jays typically remain as nonbreeders for one to several years even though they apparently are physiologically capable of breeding at age one year. In this chapter we describe how the change from helper to breeder occurs. We emphasize that dispersal characteristics differ between males and females and that the overall dispersal system is organized around an apparent dilemma, namely that nonbreeders must live and eventually breed within a habitat that is continually saturated with permanent territories.

7.1. OBTAINING BREEDING SPACE

Helpers become breeders by obtaining breeding space, which requires obtaining a mate *and* a territory. Given that Florida Scrub Jays are monogamous with equal death rates between sexes among breeders, the number of opportunities to become breeders are equal for males and females. However, a profound difference exists between the sexes in their dispersal strategies, and the dichotomy appears to revolve around the different potential each sex has for inheriting breeding space.

Male helpers obtain space in one of four ways (figure 7.1): (1) through replacing a dead or impaired breeder away from home; (2) by occupying a segment of his family's territory (i.e., territorial budding); (3) by inheriting the natal territory following the death of both breeders or, occasionally, following the death of only the male breeder; or (4) by establishing a territory *de novo* between existing territories.

Forty-eight males with known histories became breeders within the study tract during the 1970s. Of these, 24 (50%) replaced lost breeders away from their natal territories. Occasionally during the lengthy process of territorial budding a male neighbor died, and the budding male helper replaced him while inheriting some natal ground. We include 5 such cases within the 24 replacement males. Sixteen of the 48 males (33%) inherited breeding space by typical territorial budding. Inheritance of the full natal territory accounts for 6 of the 48 cases (13%). Establishment of a territory *de novo* between existing territories accounts for the remaining 2 (4%).

Females use analogous modes for becoming breeders (figure 7.1), but they rarely inherit natal ground. Those that do not replace lost breeders pair with males who are budding or inheriting a territory. Of 41 females of known histories who became breeders within the study tract, 27 (66%) replaced lost breeders who lived outside of their natal territory. Thirteen (32%) females paired with males who were gaining space through some form of territorial inheritance. One female of known history (2% of the sample) participated in establishing a territory *de novo*. No female inherited her entire natal territory; however, three nested

139

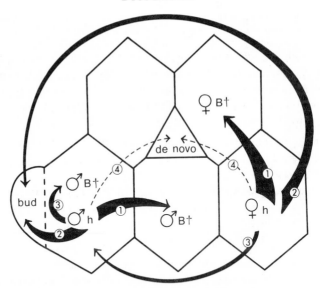

FIGURE 7.1. Florida Scrub Jays obtain space for breeding by (1) replacing dead or impaired breeders away from home, (2) territorial budding (δ) or pairing with a budded male ($\female$), (3) inheriting his natal territory (δ) or pairing with a male that inherited a territory ($\female$), or (4) establishing a territory *de novo*. Thickness of the arrows indicates frequency with which each method is used by each sex (hexagons = territories; B = breeder, h = helper, $\dagger$ = dead).

on land previously owned by the families they had helped. This occurred because these three females paired with males from adjacent territories who were gaining space through territorial budding. Their case histories are presented here as our only examples of land inheritance by females.

Case history 19. A male breeder ($-$BYSδ) died in April 1978 leaving his mate, who was incubating, and three helpers (a male, and two females including $-$PSB$\female$). The group successfully fledged his two young. In August the female breeder died and one female helper disappeared, leaving only two helpers (SGY$-\delta$, $-$PSB$\female$) and two fledglings in the 1978 territory (Appendix J, SORA). In September the two older jays acquired

140

mates from different families and portioned the 1978 territory between them (Appendix K, SORA, MIDR, respectively). As expected, the fledgling stayed with his brother. The female former helper (−PSB♀) paired with a jay (L−SG♂) who brought with him a male of unknown relationship. Though they arrived without bands, we are rather certain that both came from one unmarked family to the southeast. Therefore, in this case a female became a breeder in part of her natal territory, through the simultaneous deaths of both of her parents between successive breeding seasons.

Case history 20. In May 1971, near the end of a breeding season, the female helper (G−FS♀) of a large family isolated herself in the southern portion of the breeders' territory, while the remainder of the group, including three other helpers and two fledglings, moved northward. As this family's territory was not mapped in 1971, we do not know whether they expanded their territory, or merely concentrated their activities in the northern portion of land they owned already. By July the isolated female was chased when she attempted to rejoin her group. We do not know whether being chased or remaining apart came first. By August the female had paired with a male (A−PS♂) who had extended his activity area northward to include the place where his new mate had helped. We doubt that he won this land from her large family, but also can offer no reason for desertion of the area. In 1972 these two breeding units nested in adjacent territories (Appendix D, SCOP, COPS).

Case history 21. In 1976 two families in which the breeding males were brothers jointly defended a large territory (see case history 33; Appendix H, ECUL). Late in 1976 one of two breeding females died, and in March 1977 the widower (LWS−♂) became a replacement breeder in an adjacent territory. He left four offspring behind, including the female (FSB−♀) critical to this case. Within a month his older son (ASB−♂) joined him. His younger son (−RSW♂) and one daughter (LSB−♀) joined their uncle; in so doing both remained on natal land. The fourth offspring (FSB−♀) paired with a male from an adjacent territory (XRDS), and with much fighting they established a small territory (Appendix I, NBEN), in part *de novo* from land taken from neighbors, and in part from land once owned by her family.

141

Therefore, it is possible for females to inherit some natal ground. This apparently occurs only occasionally, when pairing with nearby males following a territorial shift that is independent of the budding event, usually associated with one or both parents' dying. In this way, these three exceptions qualitatively differ from typical budding by males, in which the parental territory remains essentially intact, except for the budded segment.

7.2. DEPARTURES OF GROUPS

The simultaneous departure of two or more helpers of the same sex from the same family is reported for several group breeders (Green Woodhoopoe: Ligon and Ligon, 1979; Acorn Woodpecker: Koenig and Pitelka, 1979; Arabian Babbler: Zahavi, 1976). This is a rare event in Florida Scrub Jays, perhaps in part because the concurrent existence in the same family of two or more helpers of the same sex is rarer than in other cooperative-breeding birds. The average number of helpers per pair with helpers is only 1.8, with the maximum number for any pair during one breeding season having been 6 (N = 1, table 5.1). From 1969 to 1979 we followed 280 seasonal breedings and found only 26 families with 2 or more female helpers (21 with 2, 3 with 3, 2 with 4) and 31 families with 2 or more male helpers (28 with 2, 3 with 3). As males help for more years than do females (figure 4.3), the greater frequency with which male helpers occur together is expected. We have two cases, both equivocal, in which two or more females may have departed simultaneously, and we have three cases for males. We present their details here as case histories, and they support our strong impression that sibs *do not move as cooperative units* between groups. In this way, dispersal by Florida Scrub Jays differs from that found in many other, apparently more "advanced" cooperative breeders.

Case history 22. In 1976, a pair nested (Appendix H, BIGL) with assistance from four helpers, all females, including three broodmates of the 1975 year class and an older stepsister (–AYS♀, see case history 6). Between censuses taken in mid-July and mid-August, all three broodmate females disappeared. Two were seen later, one (OSY–♀) in February and May 1977 about 7 km south of home and the other (PSY–♀) in June 1978 about 2 km south of home. We suppose that these three females remained together for some time after departing, but we do not know this. In any event, they clearly did not end up joining the same group.

Case history 23. In 1978, a pair nested (Appendix J, COPS) with assistance from four helpers, three broodmates (1♂, 2♀♀) of the 1977 year class and a female (–ASY♀) one year older. In January 1979 all three females were seen visiting a neighboring widowed male (SY–P♂), and two of the three were seen courting him. The older and dominant female (–ASY♀) paired with him, the "rival" sister paired with another neighbor (SGY–♂, case history 19), and the third female returned home to help one more year before dispersing to another adjacent territory and breeding in 1980.

Case history 24. In May 1975 two male helpers (YGS–♂, SR–Y♂) ejected a breeder that appeared to be sick (B–RS♂, case history 2) from an adjacent territory (Appendix G, BEND). Both cared for his fledglings and appeared to vie for the divorced female. The older and dominant male won, and in August the subordinate disappeared.

Case history 25. In summer 1976 three helper males from the same family moved into an adjacent territory (see case history 17) where at least two clearly vied for the widowed female. Again the older jay paired. After two years the paired male died and one of the other males, whom we had seen courting the again-widowed female, and who had *helped* for the two intervening years, replaced him. Only very rarely have we witnessed a helper become the mate of a jay it had helped. (The third male of this original trio, after helping them one year, budded off his own breeding space.)

A single, outstanding case of simultaneous "dispersal" is exceptional in several respects, as it involved departure of two broodmates from an apparently stable family, their adoption by a neighboring group, and the eventual pairing of one in his new territory. Details regarding the adoption are given in chapter four (case history 11).

> *Case history* 26. Two broodmates, a male (–SAY♂) and a probable male (–SFY♂) hatched in 1978, deserted their parents' territory (Appendix J, NRID) as young fledglings and moved 1 km south where they moved into a territory (Appendix J, SAND) occupied by a large family consisting of a pair, a male helper (SBW–♂), and four young of the year (2 males, 2 females). By October the fledglings and an unmarked female immigrant had joined the large family. (The female, now O–SP♀,\paired with the male helper, and they budded off a territory (Appendix K, SSAN) where they bred in 1979.)

Possibly the two adoptees were attracted to the family because the four fledglings, some occasional visitors, and the immigrant appeared to be a troop of juveniles (see section 7.6. "Fledgling Wandering"), and because of the unrest caused by the immigrant. Regardless of the cause, the adopted young of 1978 remained with this large family through early nesting in 1979 when one disappeared (and probably died). The other (–SAY♂) helped fledge young from two nests (although circumstances were unusual, in that one nestling disappeared from the first nest and two from the second). In July the two fledglings died. In August the male breeder (–ALS♂) died. In September, despite his clear, previous subordinance to the female breeder's two sons, the adopted male (–SAY♂) paired with the widow.

By October all helpers had departed, and the pair bred unassisted in 1980 and 1981.

The low mortality rate of breeding Florida Scrub Jays, which contrasts strongly with that of Green Woodhoopoes (Ligon and Ligon, 1979), may explain the rarity of simultaneous, unidirectional dispersal by the jays. Therefore, in contrast to other species such as the woodhoopoes, leaving an established group to join and help a new breeder elsewhere usually does not increase a Florida Scrub Jay helper's chances of becoming a breeder. Little cooperation is evident at this phase of their lives. Instead, dispersal is strongly an individual event for Florida Scrub Jays.

7.3. DISPERSAL DISTANCES

We estimated dispersal distances for individual jays by counting the minimum number of territories between place of helping and place of breeding (table 7.1). Because of the permanency of jay territories, the number of territories through which a jay passes to become a breeder represents a useful biological measure (Wright, 1951). Actual distance can be estimated by multiplying the number of territories by the average width of one territory (339 m).

A frequency distribution of dispersal distances by males and females is graphed in figure 7.2, again with distance measured in single territory units. Because the study tract is only about six territories wide on its longer axis (see Appendixes B–K), longer distance dispersers could be underestimated in our sample (e.g., Barrowclough, 1978). We are confident that this problem is largely circumvented by including in table 7.1 the estimated dispersal distances for marked individuals with known breeding status outside our tract. These individuals were discovered during peri-

TABLE 7.1. Florida Scrub Jay Dispersal Distances, 1970–1980

	Distance from Natal Territory[a]												
	0	.5[b]	1	2	3	4	5	6	9	12	13	15	18
Males (53)[c]													
N	7	25	10	5	4	1	0	0	0	1	0	0	0
Proportion	.132	.472	.189	.094	.075	.019	.000	.000	.000	.019	.000	.000	.000
Females (47)													
N	0	3	13	12	6	3	4	1	1	1	1	1	1
Proportion	.000	.064	.277	.255	.128	.064	.085	.021	.021	.021	.021	.021	.021

[a] Dispersal distances measured in territory widths, to nearest unit, between center of natal territory and center of breeding territory during the individual's first season as breeder. See figure 6.3 for data on drifting of territory centers through subsequent years.

[b] Nearly all cases of territorial budding are assigned a distance of .5 territory widths; centers of budded territories rarely coincide with centers of natal territories (distance = 0, as in direct territory inheritance), but are almost never as distant as a full territory width.

[c] Male sample includes one individual (SYG–δ) who dispersed two territories away, nested in 1979, then returned home and inherited his natal territory in 1980. He is tallied twice.

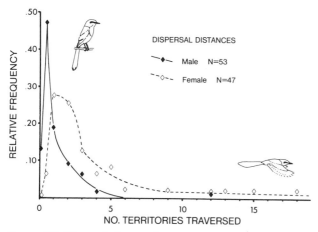

FIGURE 7.2. Dispersal distances between natal and breeding territories, for male and female jays, based on data given in table 7.1. Distance is measured in territories; the average width of one territory is 339 m. Dispersal distance of one-half indicates partial inheritance of natal territory (common for males, rare for females).

odic censuses in jay habitat surrounding the study tract (chapter two).

Figure 7.2 clearly demonstrates that male and female helpers follow different dispersal strategies. Of 49 male helpers known to have become breeders, 30 (61%) did so without moving entirely out of their natal territory, while only 3 of 43 (7%) female helpers became breeders wholly or partly on their natal ground (see case histories 19, 20, 21). Furthermore, 27 females (63%) paired 2 or more territories away from home, while only 10 males (20%) did so. The mean dispersal distance for males is .90 territories (304 m), with a median of zero, compared with a mean of 3.44 territories (1,163 m) and a median of 2 for females.

Despite significant differences in their dispersal distances, it is evident that neither sex typically disperses very

147

far. Ninety-six percent of the males breed within three territories of their natal ground, and 86 percent of the females within five territories of home. Relatively small dispersal distances raise the possibility of unusually high inbreeding coefficients in the Florida Scrub Jay, a topic discussed at the end of this chapter.

Additional evidence of the sedentary nature of nonbreeders is available. In 1979, after 10 years of color-ringing all young jays in the study tract, we made a careful search for marked dispersers within the surrounding 13 square km of suitable jay habitat (figure 3.3). We examined over 200 families and found only 9 banded jays, all breeders, of which 8 were females. These jays are included in the dispersal distance data in table 7.1. We suspect that this ratio of banded to unbanded jays outside the tract is at or near equilibrium, so that any additional dispersers might now be offset by deaths of previously established dispersers.

The results from banding Florida Scrub Jays just beyond the study tract support the conclusion that short-distance dispersal typifies the population. From 1967 through 1969 P. W. Westcott (1970), and O. L. Austin, Jr., banded 230 jays near Hicoria, which lies only 3.2 km south of our study tract. Excellent jay habitat extends throughout the intervening space. One jay, banded as a fledgling in August 1967, was the resident breeding female (Y–SA♀) about 2 km north of Hicoria and at the southern edge of our study tract (Appendixes G–K, FARS). Another jay of similar origin was captured near the northwest corner of our study tract (4 km from Hicoria) in February 1976; it has not been seen since. Unfortunately the sex of this jay remains unknown; however, its low weight (69.0 g) suggests it was a female (Woolfenden, 1978). These are the only two of Westcott's 230 jays known to have entered our study tract.

148

7.4. SEASONALITY OF DISPERSAL

Dispersal by helpers in search of breeding space does not occur with uniform frequency throughout the year. In figure 7.3 we plot the known instances of final departure from the natal territory by male and female helpers, along a 24-month axis representing their second and third years of life. As shown earlier (chapter four), these years encompass 82 percent of the first pairings by nonbreeders. Figure 7.3 separates those individuals that dispersed and paired *within* our study tract (above the axis) from those that disappeared from our study tract (below the axis). The jays tallied above the axis are all instances of known successful dispersal. The jays tallied below the axis include several (stars) we found later as breeders outside the tract, but many, probably most, of the others represent deaths either before, during, or after departure from the natal territory.

Dispersal reaches its highest frequency during two seasons, spring and fall (figure 7.3). Local dispersal, including mate replacement and new territory establishment, is highest at the onset of breeding in March. This pattern repeats itself in both year 2 and year 3, as shown in figure 7.3; the absolute size of the peak in year 3 is smaller because many helpers already have dispersed or died by that time. A smaller autumnal period of local dispersal is evident. Curiously, fall dispersal seems to occur only during year 2, immediately following a jay's first season of helping at the nest. Perhaps a fall wandering phase characterizes yearlings but not older helpers. The seasonal patterns by which helpers disappear entirely from our tract (figure 7.3, below axis) provide even stronger evidence for different age-related rates of wandering.

Our data on helpers that disappeared entirely from our records are difficult to interpret because we cannot separate those helpers that died without leaving home from those that died while attempting to disperse, or that succeeded at becoming breeders out of our sight. However, certain patterns in the data shown in figure 7.3 suggest that dispersal, or attempted dispersal, is the major source of these disappearances. The data further indicate additional differences between the dispersal strategies of male and female helpers, which we develop through the following five points. (1) As with the dispersers we know were successful, total disappearances of helpers show peak frequencies in the fall and the subsequent spring of year 2 of life. (2) The postbreeding peak in disappearances is dramatic: 38 helpers disappeared during the four-month period between June and October of their second year of life, as compared with only 26 disappearances during the remaining eight months in the annual cycle. (3) Of the known-sex helpers that disappeared, 43 (77%) were female; of these, 27 (63%) disappeared between June and October, while only 4 of 13 males (31%) disappeared in this period. (4) Again, as with known local dispersers, no autumnal peak in disappearances occurs during year 3. Indeed, no peak is apparent in the spring of year 3 either, despite the existence of a peak among the local dispersers. (5) Finally, the spring peak in disappearances seems slightly offset from the peak in locally successful pairings: of 31 successful dispersals within the tract between February and June, 26 (84%) occurred in March or April, corresponding with the peak months of nesting (Woolfenden, 1974; see chapter eight, figure 8.1). Of 15 disappearances from the tract over the equivalent period, only 5 (33%) occurred in March or April.

As mentioned above we have found a few successfully dispersed jays after they had "disappeared" from our tract. Probably a few other long-distance dispersers have escaped our detection. As shown below, however, we suspect that most of the disappearances represent deaths resulting from attempted long-distance dispersal. For this reason, the above observations regarding figure 7.3 seem especially important. The patterns suggest that during their second autumn of life (ages 14 to 17 mos.), helpers enter an active dispersal phase during which they search the surrounding habitat for potential mates. *This phase is particularly evident among females.* Those that survive, but are not successful, return home until spring, when again they begin dispersing. The first sign of the spring dispersal involves short-distance movements, when mates lost during the winter months are replaced, and new territories are established through budding. Some of the remaining helpers, unsuccessful at shorter-distance dispersals, then disappear through longer forays. Thereafter, any helper still surviving at home remains there and appears to search only for vacancies near home, especially during each subsequent spring. It remains unknown whether or not individual birds shift between long- and short-distance dispersal strategies.

7.5. DISPERSAL FORAYS

Dispersal for many bird species consists of a permanent departure, often of considerable distance, from the natal area (review by Greenwood, 1980). Dispersal attempts by Florida Scrub Jays are characterized by numerous, short-distance and short-duration forays out of the natal territory and into surrounding ones. These forays almost certainly function as a means of discovering vacancies in the breeding population. When not searching in this fashion, helpers dwell permanently in the relative safety of their familiar natal territories.

The behavior of jays on dispersal forays is quite different

151

from that of residents. The dispersers are silent and furtive. Although they appear to seek out active groups of jays, they merely perch peripheral to the group, with plumage compressed, appearing ready to avoid confrontation. Often they are chased away from the residents' territory. Only after several visits to the same group do they begin to interact. Often, after numerous visits extending over several weeks or months, a disperser stops visiting a particular family. It either directs its activities elsewhere or ceases making forays.

All jays in the study tract have been censused at least monthly for about 10 years. Many times, after several years of being recorded only in its helping territory, a non-breeder will be seen up to several kilometers or more away from home, only to be seen shortly thereafter back in its helping territory. *Such jays nearly always are females.*

Short-term absences of known-parentage, older helpers from the study tract provide a measure of seasonal and sexual differences in dispersal forays. These absences are reflected in the monthly census data. The sample (figure 7.3, top) consists of 23 helpers age 1 year or more that departed and then returned to their natal family. These include 7 males and 16 females, most of which were missed only one month before being seen again. However, 1 male was missed on two consecutive censuses, and 5 females were missed on three or four consecutively. Seasonality in these dispersal forays reflects the same bimodal pattern that characterizes permanent departures from home (cf. figure 7.3, top vs. bottom). Jays tend to wander from home in fall, after breeding and the peak of molt, and again in spring during the nesting season. A few wanderers successfully locate, and pair with, available breeders. Others return home to resume helping. However, as discussed earlier (see also chapter nine), we have strong evidence that many of these

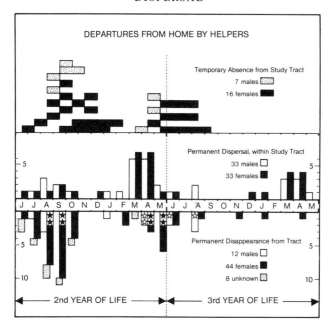

FIGURE 7.3. Monthly records of permanent (bottom 2 histograms) and temporary (upper bars) departures from home by helpers during the second and third years of their lives, when 82 percent of all first pairings occur. Two calender years, June through May, are represented along the horizontal axis. Four males (open stars) and eight females (closed stars) that moved out of the study tract permanently became known breeders; most or all of the others shown in the bottom histogram probably died. Seasonal peaks in dispersal frequency are evident in all three measures (see text).

wanderers, most of whom are females, do not survive the perils of being away from the safety of a familiar group and territory.

7.6. FLEDGLING WANDERING

In late summer and early fall, after they begin foraging independently, juvenal-plumaged fledglings frequently band together in small troops that wander through their neighborhood. These brown-headed juveniles are not chased vigorously by the resident, adult-plumaged territory holders. Although juveniles often beg from jays outside their own family, they are rarely fed. Juvenile troops typically consist of 5 to 10 jays, and the individuals that make up the troop vary from day to day. The troops disintegrate and coalesce irregularly, and the young birds typically return to their respective home territories each evening.

By October the first prebasic molt is nearly complete (Bancroft and Woolfenden, 1982), and the fledglings closely resemble adults. Also in October, resident breeders and helpers seem to defend their territories more vigorously than in summer or early fall (figure 6.1), and the troops of fledglings disband. Nearly all fledglings return permanently to their own families. Once, at this time, a pair of sibs with an extant family was adopted by another family containing four fledglings of its own (see case history 26).

Following dissolution of the early-fall troops, the young jays normally remain at home at least until after the ensuing breeding season. During 2 of the 10 years of our study we have witnessed noteworthy exceptions.

Case history 27. In March and early April of 1979, before eggs had hatched, wandering troops of from 10 to 20 non-breeders were seen; nearly all had hatched the previous year. They were seen together regularly, mostly along the edges of territories or in low-quality habitat. Several unbanded immigrants occasionally were seen with the local jays. During the late March peak of this activity, nonbreeders from as many as 10 different territories were seen together, often near the edge of a grassy swale. By mid-April the troop had disintegrated, and the nonbreeders returned home in time to feed nestlings.

Case history 28. In November 1975, after local fledgling troops had disbanded and the individuals had returned home, several fledglings from outside the tract formed a group that resided in a patch of poor habitat within the tract that consisted of rank woody shrubs and palmetto thickets bordering an isolated bayhead and associated swale (Appendixes A, H). As this habitat was not occupied by any jays, the "bay swale" immigrants confronted resident jays only when they strayed onto the surrounding oak dunes. The group was stable in size and constituency until February 1976 when it doubled to 14 individuals, including 10 jays of the 1975 year class. One of the older jays was a helper male, age 1 year 9 mo., from within the tract. The other older jays were females from outside the tract. The bay swale group remained stable through March, then began to dwindle. By May 1976, only 3 jays remained, including a 1975 year-class male (PSB–♂) apparently mated to an older, immigrant female (P–SO♀). This pair defended their poor-quality territory for two years (Appendixes I–J, NBAY) without breeding, while gradually gaining access to suitable oak habitat surrounding the swale. The female divorced and replaced a lost breeder in an adjacent territory (Appendix J, NWES) in 1978. The male acquired as a mate the female helper from this same territory and bred, finally, for the first time in 1979.

These two major occurrences of spring wandering by young jays followed the exceptionally successful breeding seasons of 1975 and 1978. Many factors may have contributed to this wandering. We suspect that successful reproduction, resulting in dense nonbreeder populations, was one important factor.

7.7. TERRITORIAL BUDDING

General Behavior

Territorial budding consists of territorial growth, followed by boundary formation between parents and son. The growth phase can be either gradual or rapid, but the boundary phase is almost always gradual. Territorial budding often is preceded by a dramatic increase in border disputes with those neighbors whose boundaries are receding. Territorial growth seems most frequent in February and March, immediately preceding the March through June nesting season. A smaller peak occurs in late summer (August–October) following breeding. Dispersal (figure 7.3) and territorial scolding (figure 6.1) also are most frequent at these times, although budding appears to occur more often during the spring dispersal season. The budding male, his parents, and to some extent any other members of the family participate in the action. One frequent result of the preliminary territorial expansion is that breeders may suddenly be found in places where they had never been seen before, displaying and calling vigorously along with the budding son.

As arguing with the neighbors subsides, often in conjunction with increased attention to nesting, the parents resume activity in their original property and the helper male remains in the newly acquired land. Pairing is often a gradual process, and several females may visit the male in his budding territory before a lasting pair bond forms. The male, before or after pairing, can return freely to his natal territory, sometimes even years after pairing. His mate, however, is excluded, primarily through chases by the resident female breeder. Time, and especially turnover among the breeders, eventually result in the solidifying of a distinct boundary between the two territories. This sequence overwhelmingly represents the commonest means by which new territories appear.

The most dramatic case of territorial budding we have witnessed caused a nest with three nestlings to fail and a pair bond to break. The sequence of events is summarized in figure 7.4. Asynchrony in the nesting cycles of two neighboring families probably helped perpetrate these events.

Case history 29. In 1973 a family discussed previously (case history 4), which included two older male helpers, had fledglings nine days out of the nest in their territory (Appendix E, COPS) when a simple pair in an adjacent territory (Appendix E, ROSE) still had nestlings, age 5 days. The larger family began fighting violently with the unassisted pair; territorial scolding was frequent and intense, and could be heard all hours of the day. Physical fighting was seen several times. After a few days the larger family occupied nearly half of the adjacent pair's territory, including land surrounding the nest. The father of the nestlings was chased when he attempted to visit his nest. The mother, though often allowed to brood, sometimes was driven away by the invading female breeder. The invading male jays, who were observed perched within .5 m of the nest, neither harassed nor fed the female or her young. The nestlings rapidly lost weight. When death by starvation seemed certain they were taken to the laboratory. The dispossessed male now wandered from the remaining segment of his territory; his mate did not accompany him. The invading breeders and one helper returned to their breeding territory. One son remained (BR–S♂), and in fall 1973, he was seen feeding the divorced female. However, by spring 1974, he had moved into an adjacent territory where he became a breeder. The second helper from the invading family (–WBS♂) paired with the divorced female and they bred successfully (Appendix F, COPB). (The divorced and dispossessed male was able to maintain a territory adjacent to his former holding, but appeared not to breed the following spring.)

In less dramatic but more typical cases, an older male helper simply may begin centering his activity at one edge of the natal territory, where he begins pairing with an immigrant female. Complete boundary formation between the new pair and the male's natal family may take months or even years. Case histories further illustrating this important dispersal event are diagramed in figures 7.5, 6.2, and previously in Woolfenden and Fitzpatrick (1978a) and Woolfenden (1976b). The most typical forms of territorial

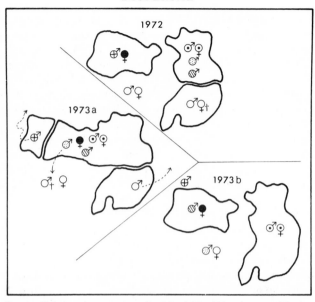

FIGURE 7.4. An extreme case of territorial budding (see case history 29 for further details). In 1972 (top), four neighboring families included one with two older male helpers (stippled, hatched). Early in 1973 (a) this large family fought violently with one of the neighboring unassisted pairs and drove off the male (crossed). This dispossessed male lost much of his territory, his mate (solid), and their brood of three nestlings. By late 1973 (b) both helper males had obtained mates, one through death of another neighbor (open). (Their parents [dots], returning to their former territory boundaries, now expanded southward through departure of a lone male there.)

budding are summarized schematically in figure 7.6, where we indicate their relative frequencies of occurrence based on the 26 cases documented through 1979.

With figure 7.6 as a model, we emphasize the continuum that exists between different types of territorial budding. Examples range from almost *de novo* territory establishment at one end to multipair territories (discussed below, section 7.8) at the other. In some cases, inheritance of portions of the natal territory accompanies simple mate replacement

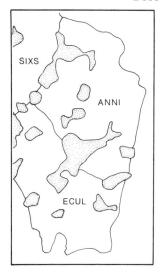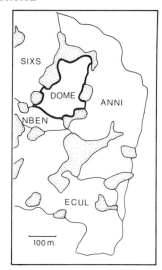

FIGURE 7.5. A classic case of territorial budding showing the boundaries before (left) and after (right) the event, which took place between the 1977 and 1978 breeding seasons. The budded territory (DOME) was established by an older helper male from ANNI territory. The stippled areas are temporary ponds, little used by the jays; three adjacent territories (SIXS, ECUL, NBEN) are labeled. (See also Appendixes I and J.)

in an adjacent territory. The important feature in all 26 cases summarized in figure 7.6 is the inheritance by male helpers of at least some land formerly (even if temporarily) occupied by his family. As mentioned above, three of these cases also involved some land inheritance by females.

Because of the importance of breeder experience to reproduction (see chapter eight), we note here that a newly budded territory virtually always contains at least one novice breeder, and usually both breeders are novices. In several cases included by us as budding, the male paired with an adjacent, widowed female who was an experienced breeder. We have no instance in which a novice female buds and pairs with an experienced male.

159

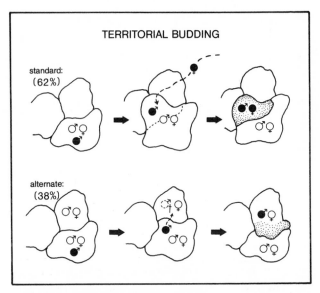

FIGURE 7.6. Territorial budding: standard and alternate methods compared schematically. Using the standard method a pair with a male helper (closed) expand their territory. On the newly gained ground, plus some possessed earlier by the family, the former helper forms a pair with a female (closed) from elsewhere, while the breeders retreat to the remnant of their former territory. With the alternate method, territorial expansion is followed by the helper (closed) replacing a lost breeder. The former helper's territory includes ground gained by mate replacement, as well as by expansion of the natal territory. Inherited ground in both cases is shown by stippling.

Frequency of Occurrence

Territorial budding is one important means by which older male helpers become breeders and is virtually the only way that new territories are created in the population. Our yearly mapping of territorial boundaries provides a measure of how common newly budded territories are in the population at large. This frequency has varied from 0 to 18 percent per year. The source of most of this variation appears to be fluctuations in reproductive success during

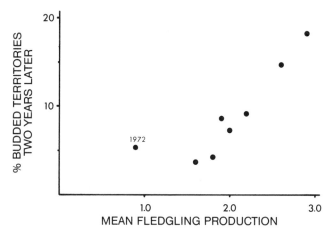

FIGURE 7.7. Correlation between fledgling production in a given year and frequency of newly budded territories two years later. The only outlier is 1972, when fledgling production was far below each of the other years.

previous years. Most males who bud off new territories do so at age 2 or 3 (see table 7.2). As shown in figure 7.7, the percentage of newly budded territories in a given year correlates strongly with fledgling production in the population two years before. Therefore, some of the dynamics of territorial compression and minor changes in breeder density discussed in chapter three can be explained as direct results of annual variation in reproductive success, these effects always buffered by a time lag of two to three years.

7.8. MULTIPAIR TERRITORIES

Through territorial budding helper jays become breeders, and a boundary forms between the former family and the new breeding pair. In the Florida Scrub Jay, then, one breeding pair per territory is the rule. In many communal breeders, however, multiple breeding pairs coexist in the

161

same territory. These include the Mexican Jay (Brown, 1963, 1970; Brown and Brown, 1981a), a close congener of the Scrub Jay.

We have witnessed five exceptions to single-pair territories during 11 field seasons. The case histories of the families involved in these five multipair territories suggest evolutionary pathways that may have been followed in the development of the more complicated communal breeding systems of other birds, and especially of other jays. Because of close filial relationships between some neighboring males, distinct territorial boundaries often are slow to develop between breeding pairs. This delay can result in a jointly defended territory containing both pairs. It seems to us that only slight alterations in the fecundity or survival of Florida Scrub Jays would increase the frequency of larger families composed of close kin. This, in turn, might increase the frequency of multipair breeding within single territories. Such an increase would result in a social system more typical of the tropical New World jays (e.g., Brown, 1963; Hardy, 1976) than of our present Florida Scrub Jay system.

One of the five cases of multipair territories resulted from a nonmonogamous mating. Woolfenden (1976b) detailed the possible causes and effects of a polygamous male who formed simultaneous pair bonds with two females (see also section 4.3. "Bigamy"). The other four cases involved closely related males. In each of three cases (case histories 30–32), a subordinate male and his mate attempted to breed within the territory of his dominant relative. In the best and longest-lasting case (case history 33), two brothers paired respectively with a widowed female and her daughter, and the four defended a common territory for two years. We emphasize this last case as most closely approximating the Mexican Jay system, both in the stable presence of two independent nesting pairs within one territory and in the

162

subsequent activities of the helpers fledged from both nests. As is true in all highly communal jays, the breeding pairs essentially confined their nest attention to their own nests. The helpers did not.

Case history 30. One male attempted to breed two consecutive years within the territory of his brother (Appendixes H, I, SORA) before securing his own breeding space. The history of the parents of these brothers probably explains their jointly defended area. The older, dominant jay (−BYS♂) helped his father for five consecutive years, the last year with a step-mother. He helped raise the jay (SY–P♂) who later nested in his territory. When the father died after breeding in 1976, the stepmother was excluded from the territory and disappeared several months later.

In 1976 the dominant male (−BYS♂) paired with a divorced breeder from an adjacent territory. His two brothers helped but the older, full brother (SY–P♂) also paired, with a one-year-old jay (PSO–♀), and they nested successfully 182 m away. The nest was well within the communally defended territory. (This provides our first and only record of a wild female Florida Scrub Jay successfully breeding at age 1 year). We suspect that this female was able to breed in the territory because her nesting was nearly synchronous with that of the dominant pair, which meant that the dominant female was concurrently occupied with incubating and brooding. In addition, her status as a yearling might have reduced the tendency for the family to drive her away during this spring wandering season. The dominant female appeared to drive the other from the territory after fledging her brood. The young female (PSO–♀) was not seen again.

In 1977 the subordinate brother (SY–P♂), with a new mate, again nested in the jointly defended territory. Their nest was concurrently active with the third nest attempt of the dominant, 132 m away. However, eggs were not seen and probably never were laid. By 1978 this pair had budded off a territory of its own (Appendix J, CABS), north of, but contiguous with, the dominant brother.

Case history 31. In 1976 a helper of three previous breeding seasons (RS–P♂) paired and nested within his parents' territory (Appendix H, XRDS) and 200 m from their nest. Three yearling sibs of the male visited both nests, and perhaps this caused the novice breeders to desert their partly incubated clutch; they abandoned their incipient territory, returned to their respective families, and helped raise those broods.

Case history 32. In 1972 a male (P–YS♂) and his helper of the two previous years (F–RS♂), with their respective mates, both attempted renesting only 30 m apart (Appendix D, CABS, CABE), F–RS♂ having paired with a widowed breeder from an adjacent territory. Although a distinct boundary had previously existed between these two territories, the line now appeared to exist only for the females. No helpers were present and neither pair fledged young. One breeder from each pair died before the 1973 breeding season, and the close association broke down.

Case history 33. The most integrated case of a multinest territory is based on the history of two brothers and a mother and daughter, all of whom came to occupy a recently vacated territory (figure 7.8). During the 1974 breeding season a female breeder (F–BS♀) was killed on the nest, and the widowed male breeder (W–PS♂) began pairing with a helper female (S–RR♀) from nearby. This helper female's father died, and her mother (P–BS♀) began joining her daughter in visits to the territory of the widowed male. Later he died, after which two old helper males from an adjacent territory incorporated the now vacant territory into a jointly defended territorial bud. The older male (G–YS♂), who was at least six years old and presumably had not yet bred, paired with the mother; the younger male (LWS–♂), who had helped for three years, paired with her daughter. Together they enlarged and defended a territory (Appendix G, ECUL) in which both pairs nested successfully 292 m apart in 1975. In 1976, again in their jointly defended territory, the two pairs nested 304 m apart. The young of the previous year, three from one pair (LWS–♂ × S–RR♀) and one from the other (G–YS♂ × P–BS♀), brought food to both nests and assisted with care of all four fledglings. We never saw the parental owners of either nest visit the other one, although they actively attended their own nests.

Following the death of his mate in 1976, the younger male

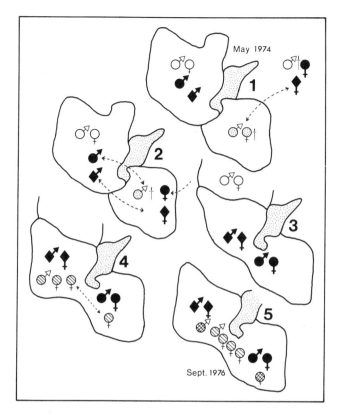

FIGURE 7.8. The formation of a multipair territory (case history 33). (1) In May 1974 a female breeder (stippled circle) died, and a female helper (closed diamond) from nearby made dispersal forays into the territory. Soon after, the helper's father (open circle) died. (2) The helper and her widowed mother (closed circle) both moved in with the widowed male (stippled circle). Soon after, that widowed male died, and two helper males (brothers; closed circle, closed diamond) from an adjacent territory moved in together. (3) The older brother (closed circle) paired with the widow, the younger paired with the novice. All four jays defended the boundary of their jointly held territory. (4) In 1975 one pair fledged one young, the other pair three young. The four fledglings formed one group, and each was fed by all four adults. (5) A year later, all four helped at both nests and fed the fledglings from each. Only one fledgling from each nest was alive in September 1976, when the system began to disintegrate through death of the younger female breeder. By 1977 (not shown), a boundary line was present between the older brother and the younger one plus his new mate; the latter pair bred far west of the territory shown here.

(LWS–♂) paired with a widowed breeder nearby (Appendix I, NWXR). The lone male helper, his son, went with him and paired with a second female, but no territorial boundary developed between them. The son (ASB–♂) and his mate did not nest in 1977. In 1978, when both pairs bred, an ill-defined boundary existed between them (Appendix J, NWXR, NRID). By 1979 this boundary had become well established (Appendix K, NWXR, NRID).

7.9. DISPERSAL AND BUDDING IN RELATION TO FAMILY AND TERRITORY SIZE

In an earlier paper (Woolfenden and Fitzpatrick, 1978a), we proposed that male helpers might directly benefit from helping to raise additional family members. Ultimately, the gain would be an increased probability of obtaining breeding space, either through territory enlargement and subsequent budding or through some advantage (e.g., dominance) in replacing neighboring breeders. Our hypothesis makes two explicit, and as yet untested, predictions about the demographic regime of dispersing male helpers: (1) that the budding event occurs principally after territorial growth and (2) that males have a measurably greater probability of becoming breeders when they live in larger families, especially if they are the dominant male helpers in those families. Here we examine these predictions with data on ages, territory sizes, and family constituencies of male helpers before and after they disperse.

Budding versus Replacing

Table 7.2 lists several demographic features of males who became breeders through territorial budding and through mate replacement. We include data only for males whose family and territory history represent typical cases of budding or replacing. Jays that inherited their breeding

166

territory through parental death (N = 6) or replaced a mate in combination with a territorial bud (N = 5) are excluded, as well as a few whose histories are either partially unknown or complicated by unusual events. Although the remaining samples are small, table 7.2 suggests that few demographic differences exist between males that become breeders through these two common methods of dispersal. Both samples come from similarly sized families, with similar familial growth patterns, similar sibling sex and age ratios, and similar territory sizes at birth and at year of helping.

The only noteworthy difference between the two samples of male helpers is that the territories of those who inherited breeding space through budding *exhibited a dramatic increase in size immediately preceding budding.* The average territory size for both replacers and budders as yearlings is similar to the average size for the population as a whole, between 9 and 10 ha. No increase occurs for jays that replace, while a 39 percent increase to 12.6 ha characterizes the territories of jays that bud. The increase occurs even though the average age of males that bud is slightly, though not significantly, younger than those that replace (table 7.2).

The size of newly budded territories averages only 4.5 ha, which is 35 percent of the average enlarged parental territory from which the new one is extracted. With one exception, these new territories varied from 2.1 to 6.8 ha (figure 7.9), significantly smaller than well-established territories (cf. figure 6.5). The remarkable exception was a 14-ha bud, which followed explosive growth of one territory to a size of 22.2 ha. This enlarged territory included large patches of poor habitat (Appendix I, NWCO). In no other case did a newly budded territory encompass more than 44 percent of the parental ground. On the average,

TABLE 7.2. Characteristics of Males Who Became Breeders through Territorial Budding and through Mate Replacement

	Budded ♂♂ (N = 16)			Replacement ♂♂ (N = 12)			Totals		
	Mean	S.D.	N	Mean	S.D.	N	Mean	S.D.	N
Family size									
As fledgling	3.47	± 1.06	15	3.00	± 0.91	13	3.25	± 0.98	28
As yearling	4.63	± 1.31	16	4.38	± 1.04	13	4.52	± 1.18	29
At departure	5.00	± 1.51	16	4.76	± 1.52	17	4.88	± 1.49	33
Sibling sex ratio (♂:♀)									
As fledglings & yearlings	19:23		16	18:17		14	37:40		30
At departure	16:15		16	17:15		14	33:30		30
Territory size (ha)									
As yearling	**9.1**	± 3.3	16	**10.3**	± 4.2	13	9.6	± 3.7	29
At departure[a]	**12.6**	± 4.7	16	**10.3**	± 5.2	13	11.5	± 4.8	29
Of new bud	4.5	± 2.9	16	—	—		—	—	
Age at departure (years)	2.6	± 0.7	18	3.1	± 1.1	18	2.85	± 0.9	36

NOTE: See 7.9 "Dispersal and Budding in Relation to Family and Territory Size" for restrictions on sample size (N = 33) used here.

[a] Among males that bud new territories, territory size at time of departure is significantly larger than as a yearling (p < .02) and significantly larger than comparable measure for males who replace (p < .05; see boldface numbers).

therefore, newly budded territories consist of acreage roughly equal to that gained by the family during the new breeder's period as a helper. Budding typically is preceded by rapid territorial growth. Therefore, it is difficult to discern whether budding occurs *because* of territorial growth or whether incipient budding *causes* the growth to take place. Especially because budding males usually are young, we suggest that budding occurs in those territories where surrounding families are less capable of defending against territorial incursion. Firm support for this hypothesis requires further study of the process.

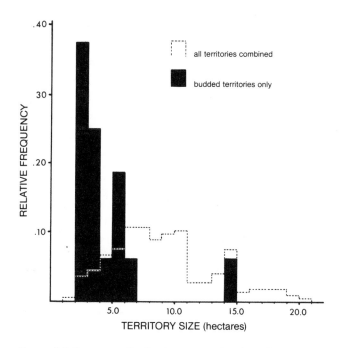

FIGURE 7.9. Frequency distributions comparing sizes of newly budded territories (N = 16) with sizes of all territories (N = 221).

Dispersal and Family Status

Table 7.2 shows that the average size of the family from which males disperse to breed, whether via budding or replacement, is almost 5. The mean is nearly as large (4.8) when all dispersed males are considered (N = 48), instead of the restricted sample used in table 7.2 (N = 33). A family size of 5 is well above average for Florida Scrub Jays at any season except after fledging (June $\bar{x}$ = 4.1). More important it is even well above the average family size (3.8) for families with helpers (whence dispersing males must originate). At the onset of breeding only 11 percent (30/280) of all families include 5 or more individuals (table 5.1; $\bar{x}$ = 3.0), suggesting the possibility that male helpers are more likely to become breeders when they come from large families.

Such a pattern supports our idea that familial growth is directly advantageous to male helpers. The critical data are presented in table 7.3, where we compare the observed proportions of males who became breeders a year after being lone helpers (.29) versus after helping with at least one sib (.46). This difference is suggestive, but it is not significant (χ^2 = 2.7; p = .10). More interesting, however, is the breakdown of the sample that had sibling helpers (N = 85). Among those males who were subordinate to another male helper, the proportion that became breeders (.31) was effectively as low as for lone helpers (.29). In contrast, dominant male helpers, be they yearlings (N = 22) or older (N = 31), became breeders with considerably higher frequency (.50 and .58, respectively). When dominant males are pooled (N = 53) and compared to subordinate and lone males combined (N = 63), their advantage in obtaining breeding status is clear and highly significant (.55 vs .30; χ^2 = 7.16, p < .01). Therefore, as

170

TABLE 7.3. Annual Probability of Achieving Breeding Status by Male Helpers of Four Social Classes

	N	Breeding Next Year	D[a]	Proportion Breeding
No sibs				
Lone helper	31	9	7	.29
With sibs				
Subordinate helper	32	10	6	.31
Dominant helper (yearling)	22	11	2	.50
Dominant helper (older)	31	18	4	.58
With sibs, pooled	85	39	12	.46
Total (all pooled)	116	48	19	.41

NOTE: Samples (N) refer to helper-years; e.g., a helper who helps alone one year and is a dominant helper the following year is tallied once as a lone helper and once as an older, dominant helper.
[a] Jays who disappeared (died or dispersed) after helping and before breeding are tallied under *D*; we strongly suspect that most or all of these males died.

best we can presently measure it, males who help raise younger sibs—and thereby dominate those sibs—have a dramatically higher probability of becoming breeders than either lone males or males who are subordinate.

7.10. EFFECTIVE POPULATION SIZE AND INBREEDING

The genetic structure of our Florida Scrub Jay population is of particular interest because of the possibility that cooperative behavior might arise through a generally elevated level of inbreeding in the population as a whole owing to reduced dispersal. Unusually high coefficients of relatedness between randomly mated individuals in a breeding unit have been hypothesized at least once in the literature on avian cooperative breeding (Brown, 1974, but see Brown and Brown, 1981a). The theoretical basis for this possibility has been explored extensively since Hamilton (1964) formalized the importance of relatedness in the evolution of so-called altruistic traits in a population (Hamilton, 1972; Maynard Smith, 1971; Michod, 1979; reviews in Markl, 1980). To date, however, no correct, data-based analysis has been carried out to examine inbreeding coefficients in a cooperative breeder. Indeed, the first analysis of this question in any bird population appeared only recently (Barrowclough, 1980). The question is additionally appropriate here because of the remarkably sedentary dispersal characteristics in our population (table 7.1), a feature that could increase the average relatedness between individuals within the population.

Our data on dispersal distances, adult longevity, and adult movement patterns allow us to model the genetic population structure of the Florida Scrub Jay. We use the continuous population model of genetic isolation by distance, originated by Wright (1943), and follow the calculation methods employed by Barrowclough (1980). In brief, the model requires calculation of neighborhood size based on the variance in dispersal distances exhibited in the population. Neighborhood size, N, is converted to effective population size, N_e, following Emigh and Pollak (1979). Finally, Wright's (1951) F_{ST} statistic is estimated from N_e, producing an estimated index of inbreeding resulting from the limits to realized population size imposed by localized dispersal. F_{ST} values in the order of .1 or higher would imply unusually high levels of inbreeding (Wright, 1978), raising the possibility that cooperation between seemingly unrelated individuals might originate through a generally elevated degree of relatedness within local demes.

We begin with Wright's (1946, quoted from Barrowclough, 1980) equation for neighborhood size, N, estimated as the number of breeding individuals occupying a circle of radius 2σ, where σ is the root mean square of dispersal distances (calculated below):

$$N = p\cdot\pi\cdot(2\sigma)^2$$

where p is the population density, measured in appropriate units. We measured dispersal distances for our jays in average territory widths (table 7.1), so $p = 2$ breeding individuals per territory in our calculations.

Calculation of σ^2 (mean square dispersal distance) typically requires summing the variance over the average lives of individuals, but this problem is simplified here because dispersal in the Florida Scrub Jay is essentially a one-time event. The relatively minor drifting of territory centers (figure 6.3) and rarity of movement by breeders to adjacent territories (section 4.5, "Fates of Widowed Jays") allow us to drop the negligible component of breeder dispersal without affecting the results. Males and females show different dispersal distances (figure 7.2), so we average their mean squares to obtain a single estimated σ^2 by the formula $\sigma^2 = (\sigma^2 \male \sigma^2 \female)/2$. Measured in territory units,

$$\sigma^2 = (4.38 \male + 26.44 \female)/2 = 15.41$$
$$\sigma = 3.93$$

The averaging, admittedly an approximation, seems justified, because in our monogamous population with equal death rate among breeders of both sexes, exactly as many males disperse as females. Therefore, the components to genetic isolation by distance contributed by each sex should be weighted equally.

Returning to Wright's equation, and substituting our values, we obtain an estimate of genetic neighborhood size

173

$$N = 2 \cdot \pi \cdot (2 \cdot 3.93)^2 = 388 \text{ individuals}$$

Again following Barrowclough (1980), we use the method provided by Emigh and Pollak (1979) for calculating effective population size, N_e, in diploid populations given overlapping generations. We use the data presented in our life table (chapter nine, table 9.10), assuming equal sex ratio at birth, a maximum life-span of 20 years, and a breeding age structure as shown in table 9.8 (assuming exponential mortality of breeders after age 10 years). In the notation of Emigh and Pollak (1979), we obtain the following three equations:

$$N_e = 5.581 N_e(1)$$
$$N_e(1) = 2 \cdot N^f(1)$$
$$N = 14.56 N^f(1)$$

where $N_e(1)$ is effective number of newborn individuals and $N^f(1)$ is number of female births each year. The figure 14.56 is exactly twice the average female lifetime. Solving the above equations, and incorporating our estimate for neighborhood size, N, yields,

$$N_e = .767N$$
$$N = 388$$
$$N_e = 298 \text{ individuals}$$

This figure is reduced slightly (to about 290) if we allow for slightly longer generation time and average lifetime among males than females.

Whereas this effective population size apparently is small for a bird (see comparative data in Barrowclough, 1980), it is by no means in the range where inbreeding effects become severe. Relative to a "universe" of one million demes, an effective population size of 290–300 yields a coefficient of inbreeding, F_{ST}, of about .025 (Wright,

1951; but see Felsentein, 1975, regarding problems with such estimations based on simple isolation by distance models).

7.11. CONCLUSIONS

Dispersal by Florida Scrub Jays is *not* a cooperative event. More jays always exist who are ready to breed than there are breeding slots open for them. When a breeding vacancy arises in the population through death of a breeder, several individuals of the appropriate sex almost invariably vie for the opportunity to pair with the widowed jay. When a male begins to establish his own territory through budding, various wandering females usually visit before he begins to pair with one of them. Older dominant helpers clearly win out over younger ones in most such competitions. Older male helpers preferentially inherit natal land even when numerous other jays are available to breed there. Apparently jays from larger families have a competitive advantage over jays from smaller ones, perhaps through subtle dominance relationships, perhaps through other advantages inherent in possessing larger territories. Dispersal is a solitary event, and sibs compete against sibs rather than joining them in dispersal attempts.

The dilemma of dispersal, especially in a densely crowded environment, is that leaving the familiar safety of home carries with it considerable risk to survival; yet some form of exodus from home is necessary if individuals in an outcrossing population are to avoid significant levels of inbreeding. Given this dilemma, and given the observation that dispersal is a highly competitive event in our Florida Scrub Jay population, we propose here a model for the origin of the sexually dimorphic dispersal system described in this chapter.

175

Suppose that a primitive dispersal system existed in which both sexes left the natal territory soon after attaining independence, just as Scrub Jays in western North America do today. As most birds apparently show greater dispersal distances among females than males (Greenwood et al., 1979a,b; Greenwood, 1980; Shields, 1982), this might have characterized the ancestral system but is not necessary in our hypothetical ancestor. The presumed mode of pairing would be that dispersing jays of both sexes fill breeding vacancies through mate replacement or by establishing new territories in unoccupied habitat patches. Individuals would experience some relatively high probability, D_o, of successfully dispersing to become a breeder at age 1 year.

Now envisage some combination of events whereby D_o is significantly reduced. This reduction could occur through an increase in breeder survivorship, a decrease or limitation of acceptable habitat patches, or an increase in mortality of the dispersers once they leave home. This demographic perturbation would select for some modification of the dispersal regime, for two reasons. First, it would increase competition for available slots, favoring dominant or more experienced individuals over subordinate ones. Second, it would favor more sedentary behavior by the nonbreeders, assuming that breeding vacancies near home could be filled with less risk of mortality than more distant ones.

Assuming no change in reproductive rates, the new regime would be accompanied either by increased juvenile mortality, or by a buildup of prebreeders, including a few older individuals that were unable to encounter breeding vacancies at age 1 year. As the nonbreeding population increases, the competitive status of yearling jays decreases because older, more experienced, dominant jays usurp more and more of the vacancies that do arise. At this point it becomes advantageous to stay home and *delay the search* for vacancies, again for two reasons. First, with an inferior competitive ability to begin with, younger jays stand an even smaller chance than the older ones in the at-large search for vacancies. The D_o is reduced even further. Second, by remaining at home the nonbreeders can begin to practice strategies, including modes of inheritance, that will enhance their opportunity to become breeders without having to disperse far from their familiar natal ground. Furthermore, in the safety of the natal territory they can gain experience and dominance status that will improve their chances of winning a breeding vacancy when they do disperse.

The stage is set for the point of divergence between males and females, and the critical variable again is dominance. As we have shown previously (Woolfenden and Fitzpatrick, 1977), males strongly and unequivocally dominate females. For this reason, males are far more likely to be in a position to inherit breeding space simply because they can drive the females away. This advantage would be strengthened by any tendency for immigrant females, once paired, to drive away other females, including formerly resident ones, as is typical in our population.

Given that one sex now begins to disperse by budding and inheriting, close inbreeding can be avoided only if the other sex disperses farther from home. Assuming that close inbreeding is deleterious, females must develop, or simply maintain, extranatal dispersal tendencies, even given a one or two year prebreeding period at home.

Key to the evolutionary sequence just described are three assumptions. The first is that inheritance provides a better avenue for nonbreeders to increase their chances of becoming breeders compared with any strategy requiring them to leave home. This assumption is supported by our evidence that survivorship within larger groups is greater than in smaller ones, and that survivorship among dispersing females is lower than in any other adult category (see chapter nine). Furthermore, individuals can affect their own probabilities of inheriting by causing territory size to increase, thereby "fighting from the inside" in the competition for breeding space. The second assumption is that male dominance was present ancestrally, resulting in males becoming the sex more capable of earning inheritance. Despite recent claims to the contrary (Smith, 1980a,b), male dominance does seem to be the rule among passerine birds, and certainly among the Corvidae (Goodwin, 1976). Third, we assume that close inbreeding is deleterious in this typically outcrossing population and that individual selection should favor a dispersal strategy that minimizes inbreeding. The theoretical discussions of this topic are mixed (Maynard Smith, 1978; Williams, 1975). Nevertheless, we observe active avoidance of close inbreeding in our population (chapter four), and the same appears to be true among other vertebrate systems studied to date (e.g., Hoogland, 1982; Brown and Brown, 1981a; Koenig and Pitelka, 1979). Thus, the animals themselves seem to be showing us that regular inbreeding is sufficiently deleterious to be avoided even in systems where it easily could be incorporated.

Dispersal is, of course, the most critical issue facing students of most cooperative systems. Among vertebrates at least, all individuals ultimately try to move into breeding positions if they survive long enough. Therefore, we are not asking more complicated questions about sterile castes or permanent helpers. Rather, we seek evolutionary answers to the problem of why individuals habitually delay breeding and help others when their proximate goal clearly seems to be to become breeders themselves. (Presumably, such answers also will help to explain why *breeding* individuals help each other in some more advanced systems.) It is now rather widely accepted that an important feature underlying many cooperative-breeding systems, perhaps most of them, is limitation or localization of some resource, frequently habitat related (Brown, 1978a; Koenig and Pitelka, 1981; Emlen, 1982a). As is already evident from preceding discussions, we view the behavior of Florida Scrub Jay helpers as more directly related to their eventual dispersal and breeding than many other authors have recognized thus far. Because the theoretical amplification of this view incorporates details on reproduction and survivorship (see chapters eight and nine), we again postpone its integration into unified fitness analyses until chapter ten.

CHAPTER EIGHT

Reproduction

SYNOPSIS

Pairs with helpers fledge 1.5 times more young than do pairs without helpers. The advantage thus gained persists after fledging. Helpers thereby provide reproductive "insurance" in that they increase a pair's probability of fledging success during a given season and therefore its probability of having at least one helper during the following season. Pairs occupying essentially the same territories in different years tend to fledge more young with helpers than without. This suggests that helpers are more important than territory quality in determining fledgling production.

Helper constituency appears not to affect fledgling production significantly. Neither number of helpers beyond one, nor sex of helper, correlates with fledgling production. A positive correlation between fledgling production and age of helper, though probably real, is not significant.

Florida Scrub Jay helpers help eggs to reach hatching and nestlings to reach fledging. They barely help fledglings to reach independence, and they do not help independent young to become yearlings. The presence of helpers increases fledging success only early in the highly synchronized, short, early spring breeding season.

Clutch size varies from two to five eggs, and the variation probably is caused by a host of social and environmental factors. Modal clutch size is three; clutches of four and five on average produce more fledglings.

The primary cause of nesting failure is predation, which is significantly lower for pairs with helpers. Hatching failure is a minor source of loss, but, interestingly, it is rarer for pairs with helpers. Starvation of nestlings is rare and is not related to presence or absence of helpers. Annual variation in mean reproductive success is high. Of four factors potentially accounting for this variation, only one—nest predation—correlates significantly with fledgling production. The correlation is virtually perfect: the annual level of nest predation determines the mean number of young that fledge.

Preliminary data suggest that jays who help longer fledge more young once they become breeders. Countering this, however, jays who help longer also show significantly lower survival rates once they become breeders. Young jays, if they can become breeders, will add more to their inclusive fitness by breeding than by helping. Per capita reproductive success, a measure whose value we question, is equal for simple pairs and pairs with one helper; thereafter a steep decline occurs as the family unit increases in size.

In nearly all bird species exhibiting delayed breeding and cooperative nesting, the number of young produced in the average year, and even the number that survive to earliest potential breeding age, greatly exceeds the number of available openings in the breeding population (Brown, 1969, 1978a; Ricklefs, 1975). The interrelationships between fledgling production, prebreeding survival, and breeder mortality profoundly affect the evolution of strategies by which breeders produce breeding offspring, and by which offspring become breeders. In this chapter we examine the first of these three factors—the production of new young jays during the annual spring breeding season. Postfledgling survival is touched upon here, then is examined in more detail in the following chapter.

Reproduction in Florida Scrub Jays is complicated by the existence of helpers in about half the territories. As shown in an earlier, preliminary examination (Woolfenden, 1975), these nonbreeders apparently improve the reproductive output of the breeders they help. A major portion of this chapter, therefore, is devoted to examining in detail the extent to which Florida Scrub Jay helpers affect the breeding success of pairs, who usually are their own parents. We begin with a brief summary of nesting season chronology.

8.1. THE BREEDING CYCLE

The relatively synchronous breeding season of the Florida Scrub Jay extends for about 90 days. Excluding one nest found in a citrus grove adjacent to the study tract, whose clutch was laid in late February, we have found no jay nests with eggs before early March, and no nest has remained active beyond late June. The overall chronology is summarized in figure 8.1, which emphasizes that nearly all jays are actively nesting by late March and most nesting is complete by late May. For reasons still incompletely understood, the nest cycle is timed to coincide with the annual, synchronous leaf drop by the common oak species, so that incubation typically takes place during a period when the nest is maximally exposed. The new flush of catkins and leaves, probably accompanied by an increase in densities of certain insect prey species, occurs soon thereafter when most pairs have nestlings or dependent fledglings.

The breeding season terminates sooner than that of many Florida land birds (Woolfenden and Rohwer, 1969; Woolfenden, 1974). Possibly the necessity of completing prebasic molt before the onset of acorn harvesting (late August) and the autumn peak in territorial defense (figure 6.1) causes the abrupt termination of nesting in June (see also section 8.4). For further speculations on Florida Scrub Jay nesting

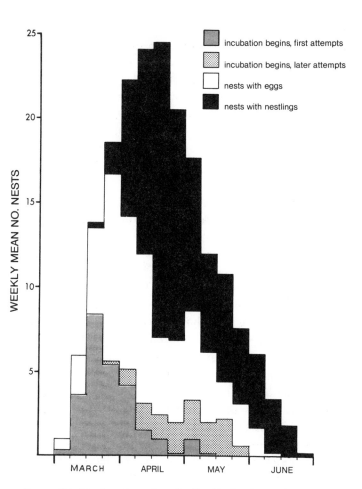

Figure 8.1. Nesting chronology for Florida Scrub Jays, based on nests found in the study tract at the Archbold Biological Station, 1970–1979. The outer line shows the number of nests with eggs or young for each week of the nesting season. Weeks of laying of first clutches (vertical hatching) and later clutches (stippling) are shown within the unshaded area representing nests with eggs. The black area represents nests with young.

in relation to rainfall and oak phenology, see Woolfenden (1974). As mentioned in chapter nine, characteristics of breeder mortality also might select for a short, early breeding season (see section 9.2. "Survival of Breeders").

Normally, a breeding pair attempts to produce only one brood of fledglings per season, but as nests often fail, the pair will renest several times in an attempt to fledge a single brood. Most pairs cease nesting after three attempts. The maximum number of attempts so far recorded is 4 (N = 1, in 1972), and the mean annual number over the entire study is 1.47 attempts per year. Among pairs that failed to fledge any young during a season, the mean number of nesting attempts containing eggs is 2.00 (N = 60).

Second Broods

True second broods are rare. Defined as a nest attempt with eggs while young from an earlier nest of the same season survive, this event has occurred only 21 times during 269 seasonal breedings monitored between 1970 and 1979. The frequency is 13 percent based on a denominator of 160 pairs that had living fledglings from first attempts of the season.

Pairs with helpers are more likely to attempt true second broods than pairs without helpers. Eighty-six percent of the pairs attempting second broods had helpers, a proportion much higher than for the population as a whole (57%). Certain pairs are likely to be repeaters at second brood attempts. In all, only 12 pairs account for the 21 attempts, including 1 pair that attempted second broods six consecutive seasons. One season this pair tried renesting twice after a successful first nest.

Frequency of double-brood attempts varies from year to year. Seventeen of the 21 attempts occurred during four years (1975, 1976, 1978, 1979) when 76 pairs had the opportunity to attempt second broods. The remaining 4 attempts were lone instances during four of the other seven years, when 84 pairs had the opportunity.

At present we do not know what factors stimulate second brood attempts. Regardless, the effect on the total number of young fledged (26 of 549) during the decade has been small.

8.2. ANNUAL FLEDGLING PRODUCTION

Florida Scrub Jay pairs fledge an average of about 2.0 young each year, as shown by the overall summary data in table 8.1. Mortality is high during the 2-month period of dependence, and only about 1.2 fledglings survive to be counted on the July census, our arbitrary date for reaching independence. An average of only .7 young remain at the end of 1 year postfledging. This 12-month period of declining mortality is examined more closely in chapter nine, but the overall results warrant emphasis here. For a Florida Scrub Jay to replace itself *just with a yearling* (still prebreeding age in virtually all cases), it must count on an average of three years' breeding. The variance in breeding output is high (table 8.1), and many pairs breed for several successive years, with 2 to 3 attempts each year, without succeeding in fledging a single young jay.

During 10 breeding seasons (1970–1979) we followed the breeding of 269 pairs, of which 152 (57%) were accompanied by one or more helpers. Table 8.1 shows that in these samples taken as a whole, pairs without helpers fledged an average of 1.59 young, while those with helpers fledged 2.39 young per pair. This produces a difference that persists through the year, so that by the following breeding season pairs without helpers average .57 yearlings pro-

TABLE 8.1. Mean Production of Young Florida Scrub Jays (1970–1979)

	Seasonal Breedings	Fledglings/ Pair	Independent Young/Pair	Yearlings/ Pair[a]
Without helpers	117	1.59 ± 1.25[b]	0.91 ± 1.05	0.57 ± .84
With helpers	152	2.39 ± 1.50	1.48 ± 1.22	0.77 ± .97
All pairs	269	2.04 ± 1.39	1.23 ± 1.18	0.68 ± .92

[a] Excludes 1979 data; N = 233 seasonal breedings (105 without helpers, 128 with).
[b] Figures shown are means over all seasonal breeding attempts through 10 years, ± standard deviation; all years are pooled. See table 8.2 for annual breakdown and average *annual* reproductive output.

duced versus .77 yearlings for pairs who had helpers (table 8.1). This difference means that pairs with helpers produce 35 percent more yearlings than pairs without. The observation that previous success at breeding provides helpers, who in turn dramatically increase subsequent breeding success, is of great consequence to the demography of the population. We shall examine these consequences later, but here it remains to establish more conclusively that helpers indeed *help* pairs raise significantly more offspring.

8.3. DO HELPERS HELP?

Table 8.2 summarizes 10 years' reproductive data, documenting the consistent difference between pairs with helpers versus pairs without helpers. Annual production of young is extremely variable, for reasons still incompletely understood (see section 8.6). During the most productive year (1970), fully 2.5 times as many young fledged per pair as during the worst (1972). These annual fluctuations are graphed in figure 8.2 (bottom). Throughout all 10 years of study, however, the relationship has remained that more young are produced by pairs with helpers than by those without. As shown in table 8.2, this difference is relatively constant, representing an absolute increase of .9 fledglings produced by pairs with helpers compared to unassisted pairs. As a result of this difference, pairs that have helpers more frequently succeed in fledging at least some young, even in the worst years for breeding. Figure 8.2 (top) graphs the annual percentages of pairs fledging at least one young. The pairs with helpers have a higher success rate, and their rate is less variable year to year.

Figure 8.2 (bottom) shows that good and bad years for reproduction seem to affect pairs equally, regardless of the presence or absence of helpers. This can be seen more explicitly in figure 8.3, which documents the covariance between rates of fledgling production by pairs with helpers versus pairs without helpers. The regression line in figure 8.3 again shows that the effect of having a helper is to add approximately one fledgling (y intercept ≈ 1.0) to the productivity of lone pairs regardless of the absolute rate of production in any given year. Factors causing widespread success or failure in a certain year are equally important to all pairs. Helpers in effect provide a form of reproductive insurance by both increasing the probability of success, and increasing a pair's absolute productivity when it is successful.

More Explicit Tests

The above data are sufficient to show that the presence of helpers correlates with increased reproductive success. Several possibilities exist by which this *correlation* could be obtained even if helpers themselves did not actually help the breeders. Novice breeders or consistently unsuccessful ones, for example, rarely have any offspring surviving to become helpers. Therefore, the pairs with helpers will tend

187

TABLE 8.2 Annual Production of Young Florida Scrub Jays

	Without Helpers				With Helpers				Absolute Difference	Combined		
Year	No. pairs	Fledgl./pair	Indep. yg./pair	Yearl./pair	No. pairs	Fledgl./pair	Indep.yg./pair	Yearl./pair	Fledgl./pair	Fledgl./pair	Indep.yg./pair	Yearl./pair
1970	7	2.29±1.11	1.43	1.14	7	3.57±0.53	1.71	1.43	1.3	2.93±0.84	1.57	1.29
1971	4	1.00±1.41	0.75	0.25	18	2.00±1.41	1.39	0.67	1.0	1.82±1.38	1.27	0.59
1972	13	0.31±0.85	0.08	0.08	15	1.40±1.40	1.00	0.33	1.1	0.89±1.16	0.57	0.21
1973	14	1.36±1.50	0.57	0.36	12	1.92±1.44	1.58	0.75	0.6	1.62±1.44	1.04	0.54
1974	13	1.46±1.27	0.77	0.38	15	2.13±1.60	1.20	0.53	0.7	1.82±1.43	1.00	0.46
1975	16	2.31±1.20	1.44	1.06	11	3.00±1.73	1.73	1.27	0.7	2.59±1.41	1.74	1.15
1976	11	0.91±1.22	0.55	0.27	19	2.42±1.50	1.16	0.47	1.5	1.87±1.38	0.93	0.40
1977	13	1.85±1.21	1.15	0.62	13	2.62±1.19	1.85	1.23	0.8	2.23±1.18	1.50	0.92
1978	14	2.00±1.57	1.36	0.86	18	2.50±1.72	1.44	0.89	0.5	2.28±1.63	1.41	0.88
1979	12	2.08±1.44	0.91	—	24	2.83±1.83	1.48	—	0.8	2.58±1.69	1.56	—
Mean annual production		1.56	0.90	0.56[a]		2.44	1.45	0.84[a]	0.9	2.06	1.26	0.72[a]
±1 S.D.		0.67	0.45	0.38[a]		0.62	0.27	0.39[a]		0.59	0.36	0.36[a]
Mean production	117	1.59	0.91	0.57[a]	152	2.39	1.48	0.77[a]		2.04	1.23	0.68[a]

[a] Excludes 1979 young; see Epilogue.

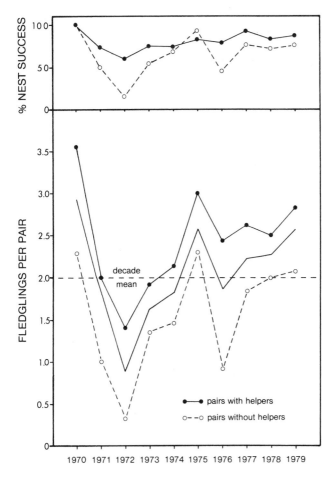

FIGURE 8.2. Nesting success (top) and fledgling production (bottom), 1970–1979, for pairs with helpers (closed circles) and pairs without helpers (open circles). The upper graph shows that fledging at least one young has been more frequent most years and less variable from year to year for pairs with helpers. The lower graph shows that the number of fledglings produced per pair has been higher each year for pairs with helpers. Intermediate line in lower graph shows fledgling productions for pooled population.

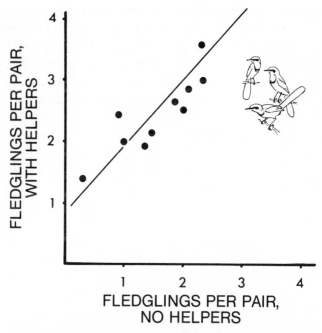

FIGURE 8.3. Correlation between rates of fledgling production for pairs with and without helpers. Good and bad years for reproduction affect both kinds of pairs equally; $r^2 = .76$.

to be more experienced or inherently more successful breeders than pairs lacking helpers, thereby resulting in higher birth rates among the sample with helpers. This difference even could be reflected in different numbers of nesting attempts, or numbers of eggs laid in each nest, between the two samples.

Tables 8.3 and 8.4 address the possible effects of breeder experience by separating breeding pairs into three samples: both novices, both experienced, and mixed. In these analyses a novice is defined as a jay that never before has tended its own nest with eggs, and a mixed pair represents a mating between a novice and an experienced breeder. Under these categories we compare mean number of nesting attempts and mean clutch sizes per year for pairs with and without helpers (table 8.3). As expected (e.g., Lack, 1968), a slight increase in both measures occurs from novice pairs through mixed pairs to fully experienced pairs. However, no difference exists *within* these categories for pairs with and without helpers. Using the same three classes of breeding pairs, and again controlling for presence or absence of helpers, we next examine fledging success (table 8.4). As with egg production, novice breeders tend to be less productive at fledging young. However, the important differences again emerge: regardless of their experience, and despite their beginning with equal numbers of eggs (table 8.3), pairs with helpers fledge more young than pairs without. The more general question of how average fecundity varies with breeder age is discussed later, in section 8.7.

Another possible bias with these measures involves differences in territory quality, which could affect reproductive success consistently enough to produce the above results. Brown and Brown (1981b) concluded that vegetational characteristics of different territories correlated with fledgling production in the Grey-crowned Babbler. (By reducing numerous babbler groups to equal size, they also concluded that helpers did indeed increase fledgling production.) Other studies support these conclusions regarding the importance of variation in territory quality to reproductive success among cooperative breeders (Brown and Balda, 1977; Gaston and Perrins, 1974).

For numerous reasons, we have chosen not to conduct experimental, controlled removals of jays in our tract. However, the duration of our study allows us to make analyses that we think control for any effects territory quality might have on reproduction. We present two of these here, both

191

TABLE 8.3. Nesting Attempts and Clutch Size as Related to Breeding Experience of Pairs

	Pairs without Helpers			Pairs with Helpers		
	No. pairs	Mean no. attempts	Mean clutch size	No. pairs	Mean no. attempts	Mean clutch size
Both novice	17	1.24	3.05 ± .50	5	1.20	3.00 ± .00
Mixed[a]	35	1.47	3.13 ± .52	30	1.48	3.30 ± .68
Both experienced	64	1.53	3.46 ± .83[b]	118	1.50	3.43 ± .68[c]
Total[d]	116	1.47	3.31 ± .71	153	1.49	3.39 ± .67

[a] No difference between novice ♂ and novice ♀ mates; therefore samples pooled.
[b] Mean clutch size of experienced pairs without helpers significantly larger than that of novice pairs ($t = 2.18$, $p < .05$), mixed pairs ($t = 2.59$, $p < .02$), and the pooled sample ($t = 3.15$, $p < .01$).
[c] Differences between experienced and novice or mixed pairs not statistically significant ($t = 1.65$, $p < .10$).
[d] No significant differences between any comparison made across the table, i.e., between analogous samples with versus without helpers.

TABLE 8.4. Fledging Success as Related to Breeding Experience of the Pair

	Pairs without Helpers			Pairs with Helpers[a]			All Pairs		
	No. pairs	Fledgl./ pair	Prop. successful	No. pairs	Fledgl./ pair	Prop. successful	No. pairs	Fledgl./ pair	Prop. successful
Both novices	17	1.24 ± 1.39	.47	5	2.20 ± 1.30	.80	22	1.45 ± 1.41	.55
Mixed[b]	35	1.34 ± 1.21	.63	30	2.47 ± 1.94**	.77	65	1.86 ± 1.60	.69
Both experienced	64	1.80 ± 1.25[c]	.73	118	2.38 ± 1.37**	.81	182	2.17 ± 1.31[c]	.78
Total	116	1.58 ± 1.25	.66	153	2.39 ± 1.50***	.80*	269	2.04 ± 1.39	.74

[a] Pairs with helpers significantly more successful than pairs without:*, $p < .05$;**, $p < .01$;***, $p < .001$.
[b] No difference between novice ♂ and ♀ mates; therefore samples pooled.
[c] Experienced pairs show significantly higher fledgling production than pooled samples containing novices ($p < .05$) except in the case of pairs with helpers.

of which further indicate that helpers themselves contribute to the reproductive success of the pairs they help.

First, we measure fledgling production by breeders who were living in essentially the same territories during years with, and years without, helpers (table 8.5). Our territory maps (Appendixes B–K) show that these territories occupy essentially the same patches of scrub year after year, even when their ownership changes. The assumption here is that any annual differences that might exist in territory quality do not affect the analysis, because the samples combine the years.

In order to qualify further for our test, the resident breeders must have attempted reproduction at least two years without helpers and two years with helpers. This restriction reduces statistical noise caused by single-year samples. During the 5 to 11 years each territory was studied, replacement breeders moved in, but at a slow rate; at least one member of the pair was an experienced breeder during every year in our sample. Using these criteria, pairs occupying the same territories fledged more young when helpers were present than when helpers were absent (table 8.5). The differences are significant in mean number of fledglings per pair (2.40 vs 1.65; $p < .01$, Student's t) and in the trends displayed within individual territories between years with and without helpers (Wilcoxon test; $T = 28.5$, $p < .01$).

The most restrictive test we can devise for examining the effect of helpers on birth rate uses only *enduring* pairs that *have* previously fledged young *and* that also bred on the same territory in later years both with and without helpers. In this test, the pairs are held constant, the territories are held roughly constant, and degrees of experience and demonstrated success are constant. Thus the major variable is presence or absence of helpers. In this stringent test (table 8.6), again pairs with helpers fledged more young

194

ᴀʙʟᴇ 8.5. Fledgling Production in "Same Territory" during Years without and th Helpers

rritory Name	No. years without Helpers	No. Fledglings	Mean no. Fledglings	No. years with Helpers	No. Fledglings	Mean no. Fledglings	Differ-ence
ᴀZ	2	3	1.50	9	29	3.22	+1.72
ʟA	4	6	1.50	5	16	3.20	+1.70
TRE	6	8	1.33	2	6	3.00	+1.67
RCT	3	4	1.33	2	6	3.00	+1.67
NNI	5	6	1.20	4	11	2.75	+1.55
ᴀND	3	5	1.67	4	12	3.00	+1.33
ᴐSE	5	6	1.20	4	10	2.50	+1.30
ᴀBS	4	7	1.75	5	15	3.00	+1.25
RE	3	2	0.67	6	11	1.83	+1.16
ᴵN	2	2	1.00	5	9	1.80	+0.80
LIN	3	9	3.00	4	9	2.25	−0.75
ᴸOT	5	10	2.00	3	4	1.33	−0.67
ᴐWL	3	1	0.33	4	4	1.00	+0.67
CUL	2	6	3.00	7	17	2.43	−0.57
ᴺND	4	9	2.25	6	16	2.67	+0.42
GL	2	3	1.50	7	13	1.86	+0.36
RID	3	8	2.67	3	7	2.33	−0.34
ᴄUL	3	7	2.33	7	14	2.00	−0.33
ᴐtals ᴅ.	62	102	1.65 1.40	87	209	2.40[a] 1.49	

= 3.10, $p < .01$; $T = 28.5$, $p < .01$ one-tailed Wilcoxon matched pairs sign-rank t.

(2.27 vs 1.84), but the difference is not significant. Deleting the requirement that pairs must have fledged young in order to be included doubles the small sample size, and produces similar (though still not significant) results: when pairs have helpers they produce more young than they do without helpers (table 8.7). The trend in all the data thus far is clear: helpers increase reproductive output, and hence the fitness, of the breeding pairs they help.

Effects of Helper Constituency

The numbers, sexes, and ages of helpers vary considerably from family to family. Having shown that the pres-

TABLE 8.6. Fledgling Production by Same Pairs, All Previously Successful at Fledging Young in "Same Territory," during Years without and with Helpers

	Without Helpers	With Helpers
Seasonal breedings	19	30
Mean	1.84	2.27[a]
S.D.	1.43	1.62

[a] $t = .95$, difference not significant.

ence of helpers positively affects reproduction, we briefly examine the effects of variation in helper constituency on the reproductive performance of the pair. Tables 8.8, 8.9, and 8.10 summarize the pertinent data.

Between 1970 and 1979 we observed the nesting of 152 pairs with helpers (see table 8.1). As detailed in chapter five, the mean number of helpers for pairs with helpers has been 1.8 helpers per pair. Only 12 pairs have had more than 3 helpers, and these are pooled in our analysis (table 8.8). The results are unambiguous: among pairs with helpers, fledgling production shows *no increase whatsoever* as helper number increases. Table 8.8 shows that the major difference lies between 0 and 1 helper; pairs with 2 or more helpers show exactly the same rate of fledgling production as pairs with only one.

TABLE 8.7. Fledgling Production by Same Pairs in "Same Territory" during Years without and with Helpers

	Without Helpers	With Helpers
Seasonal breedings	39	66
Mean	2.18	2.52[a]
S.D.	1.30	1.61

[a] $t = 1.12$, difference not significant.

TABLE 8.8. Number of Helpers and Annual Fledgling Production

	Number of Helpers				
	0	1	2	3	>3
No. of families	117	74	47	19	12
Mean no. of fledglings	1.59	2.45	2.30	2.42	2.42
S.D.	1.25	1.55	1.37	1.95	1.00

To examine as precisely as possible the effects of sex of helper on fledgling production, we separated the samples by age and number (table 8.9). These data show no difference by sex, but they do indicate a trend toward greater fledgling production with older helpers present. Breeders fledge equal numbers of young when assisted by one or more helpers of either sex when the male–female comparison is made within any age class.

Having found no difference in fledging success associated with sex of helpers, we ignore this variable by pooling the sexes to examine the effects of helpers' ages. The difference between fledging success with versus without an older helper (2.83 vs 2.20 for single-helper pairs) is suggestive, but not significant (table 8.10). As with virtually all previous analyses of fledgling production, the variance is extremely high. Larger sample sizes may eventually prove that older helpers help more than young ones. Such a result would parallel the findings of Stallcup and Woolfenden (1978) regarding the increased attentiveness of older, more experienced helpers. Lawton and Guindon (1981) found a similar trend in the Brown Jay, where helper experience does significantly affect fledgling production. At present, however, our data on Florida Scrub Jay helpers remain equivocal on this issue.

197

TABLE 8.9. Sex of Helper(s) and Annual Fledgling Production

| | | Pairs with Helpers | |
		♂ helper(s)	♀ helper(s)
1 older[a]	N	21	11
helper	$\bar{x}$	2.76	3.00
only	S.D.	2.03	0.78
1 yearling	N	21	17
helper	$\bar{x}$	2.14	2.24
only	S.D.	1.74	1.44
More	N	15	11
than one	$\bar{x}$	2.00	2.00
helper	S.D.	1.36	1.67
Yearling	N	57	39
and older[a]	$\bar{x}$	2.33	2.39
helpers	S.D.	1.78	1.39
combined			

[a] Age 2 years or older.

TABLE 8.10. Age of Helper(s) and Annual Fledgling Production

| | | Pairs with Helpers | |
		Older helper(s) present	Older helper(s) absent
One helper	N	99	88
only[a]	$\bar{x}$	2.83	2.20
	S.D.	1.64	1.56
Any number	N	80	70
of helpers	$\bar{x}$	2.55	2.29
>1	S.D.	1.56	1.63

[a] Mann Whitney U, $Z = 1.20$, $p = .11$, n.s.

Survival of Eggs, Nestlings, and Fledglings

For most bird species, it is nearly impossible to follow young birds once they leave the nest. For this reason, studies of avian reproduction generally concentrate on survival rates and nesting success only through the fledging of young (Ricklefs, 1973). Typically, it is assumed that the number of offspring surviving to independence or to reproductive age is directly proportional to the number of fledglings produced (Lack, 1966). Because Florida Scrub Jays are so sedentary, our study allows us to examine the survival of young from the time the egg is laid through at least age 1 year. Furthermore, having shown that helpers do increase reproductive success of pairs, we can examine each stage of the reproductive cycle independently to discover where, and perhaps even how, helpers influence reproduction.

Table 8.11 summarizes 10 years' data on survival of young jays. Survival ratios are shown and are separated according to presence or absence of helpers, for each of four developmental stages: eggs, nestlings, fledglings to independence, and independent young to yearlings. Our measure of egg survival in this analysis includes eggs lost before a complete clutch was laid (see table 8.13 for data that exclude these 18 eggs).

The comparisons in table 8.11 show a striking change in the effects of helpers as the young develop and mature. The positive influence of helpers on both egg *and* nestling survival is highly significant. Because the benefits to these two stages are additive, each egg laid by a female with helpers stands fully 44 percent greater probability of surviving to fledge than an egg laid by a female without helpers (.471 vs .328 probability of fledging, respectively). Once the young have fledged, however, the influence of helpers on survival begins to decline. The greater survival of fledglings

TABLE 8.11. Survival Ratios for Four Stages of Florida Scrub Jay Reproduction, 1969–1979

	Nestlings/ Eggs		Fledglings/ Nestlings		Indep. Young/ Fledglings		Yearlings/[a] Indep. Young	
	No.	Ratio	No.	Ratio	No.	Ratio	No.	Rati
Without helpers	$\dfrac{342}{567}$	.603	$\dfrac{186}{342}$	.544	$\dfrac{105}{186}$	.565	$\dfrac{60}{95}$	.63!
With helpers	$\dfrac{531}{770}$	.690	$\dfrac{363}{531}$	.684	$\dfrac{225}{363}$	.620	$\dfrac{99}{180}$	.55!
χ^2	10.77***[b]		17.41***		1.57[c]			1.7
Total	$\dfrac{873}{1337}$	.653	$\dfrac{549}{873}$	.629	$\dfrac{330}{549}$	.601	$\dfrac{159}{275}$	.57

NOTE: Sample pools all nesting attempts, including clutches preyed upon during laying a very few nests found with nestlings for which clutch sizes were estimated.
[a] Data on yearling survival excludes 1979 sample.
[b] χ^2 reflects significance of difference between survival ratios without and with helper $p<.001$.
[c] Not significant.

to independence with helpers (.620 vs .565) is not significant, although it may be biologically real. The trend actually may be reversed between independence and age 1 year (.550 survival with helpers vs .632 without), but again the difference is not significant.

The most frequent cause of egg and nestling mortality is predation of complete clutches and complete broods (see 8.4 "Causes of Nesting Failure"). Therefore, the fate of an egg or nestling is not statistically independent of its nest mates. This situation violates slightly the assumption of independence implicit in our use of the Chi-square tests in table 8.11. More legitimate is to compare the performances of pairs with and without helpers for proportion of clutches that survive to produce nestlings (.781 vs .719, respectively) and proportion of broods that survive to produce fledglings (.747 vs .642, respectively). The difference in probability of producing nestlings is not significant ($\chi^2 = 2.0$, $p > .05$); the difference for fledglings is significant ($\chi^2 = 3.84$, $p = .05$). For pairs with helpers survival of clutches was greater in seven of the ten years (n.s., Wilcoxon sign-rank test), and survival of nestlings was greater in nine of the ten years ($p < .05$). Finally, the chances of a clutch surviving to produce at least one fledgling are significantly greater for pairs with helpers than pairs without (.583 vs .462, $\chi^2 = 5.78$, $p < .02$).

Whether or not the different postfledging survival rates shown in table 8.11 are biologically significant remains moot. Based on the data thus far available, however, the net probability of a *fledgling* surviving to age 1 year is virtually identical between the helper versus no helper samples (.341 vs .357, respectively). This result is graphically illustrated in figure 8.4, where the number of independent young and yearlings produced per pair are graphed against mean fledgling production over the ten years studied. The sample with helpers consistently produced more independent young and yearlings, but only because they had produced *more fledglings to begin with*. The two samples in each graph follow the same regression line, which represents a shared annual survival probability of .35 (see chapter nine).

Fledgling survival could be related in some way to the number of sibs fledged from the same nest. Intrafamilial competition might reduce survival in larger broods; conversely, cooperative effects within larger broods might increase survival relative to smaller broods. Table 8.12 addresses this question and shows that survival to age 1 year is independent of fledgling brood size. The mean number of fledglings that survived to become yearlings from each group size matches the number predicted to

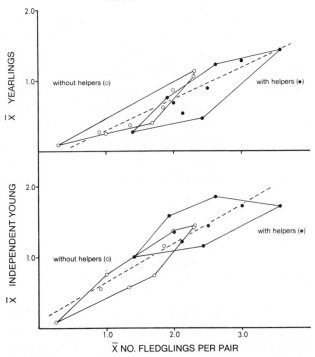

FIGURE 8.4. Correlation between mean number of fledglings produced annually by pairs with (closed circles) and without (open circles) helpers and mean number of independent young (age approximately 3 months) and yearlings (age approximately 14 months) they produce. Pairs with helpers consistently produce more independent young and yearlings, but only because they produce more fledglings (note that with-helper and without-helper samples fall along the same regression line).

have them, based on 35 percent survival, for each category of fledgling brood size. Thus, the number of fledglings produced by a given pair in one year does not affect the survival probabilities of those fledglings. Further analyses of postfledging survival, and discussion of its implications, can be found in our broad treatment of survivorship (see chapter nine).

TABLE 8.12. Survival of Different-Sized Fledgling Groups

	No. of Fledglings during One Season				
	1	2	3	4	5
No. of families	20	49	68	24	5
No. of yearlings	0.35	0.63	1.19	1.38	1.40
No. predicted (35% survival)	0.35	0.70	1.05	1.40	1.75
Proportion with ≥1 yearling	0.35	0.53	0.71	0.75	0.60
Proportion predicted (35% survival)	0.35	0.58	0.72	0.82	0.88

Because postfledging survival appears not to be affected by helpers, the fact that pairs with helpers succeed in fledging half again as many young as do those without helpers represents a fundamentally important feature of Florida Scrub Jay reproduction. We now turn to a closer examination of where in the nest cycle the helpers make such an important contribution, our ultimate goal being to discover *how* helpers help.

Survival through the Nest Cycle

To examine the effects of helping as the season progresses and to provide more details regarding survival through the nest cycle, we calculated survival rates throughout both the egg and nestling stages, separating *first* nest attempts (N = 256) from *renests* following a first attempt's failure (N = 109) or fledging (N = 21). The median date of beginning incubation for the later attempts (3 May) is about 40 days later in the season than that of first attempts (25 March). In these calculations we use only completed clutches that were monitored from near the beginning of incubation (sample size of eggs reduced from 1,337 in table

8.11 to 1,294 here). The data and survival rates are shown in tables 8.13 and 8.14, and they demonstrate three new features of Florida Scrub Jay breeding.

1. As has been shown for some other bird species (e.g., Lack, 1954, 1966), eggs and nestlings from seasonal first nest attempts are more likely to survive than those from later attempts. Pooling the helper and no helper samples gives the following results: eggs, .828 versus .670 (χ^2 = 41.3, p < .001); nestlings, .696 versus .448 (χ^2 = 45.6, p < .001).

2. Survival rates do not remain constant through the nest cycle, but tend to decline with time in the nest (see Ricklefs, 1969; Caccamise, 1976). Particularly toward the end of the nestling period (table 8.14), the probability of a living young surviving through each subsequent time period declines. This measure often is assumed to remain constant at least throughout incubation (e.g., Mayfield, 1961; see Johnson, 1979), but even during this period the daily survival rate appears to decline significantly in all but the sample of first nest attempts with helpers (table 8.13). Stated another way, the longer an egg or nestling lives, the greater is its daily probability of dying.

3. Perhaps most surprising, the positive effects on the survival of both eggs and nestlings by helpers occur *only* during the season's first nest attempt. Later nests show statistically equivalent survival rates, during both nest stages, between samples with and without helpers. Whatever is the mechanism by which helpers help, we have now narrowed the event to some feature that occurs early in the nesting season, but is either absent or swamped during later attempts.

Figure 8.5 plots the cumulative survivorship of young jays, from initial incubation of eggs through fledging of young, on the customary semilogarithmic axes. A constant

TABLE 8.13. Survival of Eggs during Incubation for Seasonal First Nests and Renests for Florida Scrub Jay Pairs without and with Helpers

	Days of Incubation									
	Seasonal first nests					Seasonal renests				
	0	4	8	12	17[a]	0	4	8	12	17[a]
Without helpers										
No. eggs	361	352	337	327	285	174	163	149	135	114
Prop. surviving	1.000	.975	.934	.906	**.789**	1.000	.937	.856	.776	**.655**
Survival rate	—	.975	.957	.970	.872	—	.937	.914	.906	.844
With helpers										
No. eggs	506	497	478	451	433	253	246	224	207	172
Prop. surviving	1.000	.982	.945	.891	**.856***	1.000	.972	.885	.818	**.680**
Survival rate	—	.982	.962	.944	.960	—	.972	.911	.924	.831

[a] Hatching occurs after about 17 days.
*Proportion of eggs surviving to hatching is significantly higher for pairs with helpers versus pairs without helpers for seasonal first nests (.856 versus .789, $\chi^2 = 6.50$, $p < .02$), but not for seasonal renests (.680 versus .655, $\chi^2 = .28$).

TABLE 8.14. Survival of Nestlings for Seasonal First Nests and Renests for Florida Scrub Jay Pairs without and with Helpers

	Days of Nestling Life									
	Seasonal first nests					Seasonal renests				
	0	4	8	12	17[a]	0	4	8	12	17[a]
Without helpers										
No. nestlings	238	217	197	172	143	91	84	64	47	40
Prop. surviving	1.000	.912	.828	.723	**.601**	1.000	.923	.703	.516	**.440**
Survival rate	—	.912	.908	.873	.831	—	.923	.762	.734	.851
With helpers										
No. nestlings	384	361	344	329	290***	148	135	117	89	67
Prop. surviving	1.000	.940	.896	.857	**.755*****	1.000	.912	.791	.601	**.453**
Survival rate	—	.940	.953	.956	.881	—	.912	.867	.761	.753

[a] Fledging occurs after about 17 days.
*** Proportion of nestlings surviving to fledging is significantly higher for pairs with helpers versus pairs without helpers for seasonal first nests (.755 versus .601, $\chi^2 = 16.55$), but not for seasonal renests (.453 versus .440, $\chi^2 = .04$).

TABLE 8.15. Fledgling Production from First and Later
Seasonal Nest Attempts

		First Nests		Renests	
	N	No. fledglings	Proportion	No. fledglings	Proportion
Without helpers	186	140	.813	46	.187
With helpers	363	295	.753	68	.247
Total	549	435	.792	114	.208

slope would indicate constant mortality rate of nest contents
through the 36 days required for fledging. The only nests
that approximate such a constant rate are first nest attempts
by pairs with helpers. As a result, this sample shows greater
survival overall than any of the other samples. The implications of these data on the fledging success of pairs with
helpers versus without helpers are further emphasized by
the data in table 8.15. Nearly 80 percent of all fledglings
produced during the breeding season are products of a
seasonal first nest attempt. Together, these data indicate
the importance of succeeding early in the season. The presence of helpers provides some measure of insurance that
a pair will realize this early success.

8.4. CAUSES OF NESTING FAILURE

To understand how helpers increase reproductive success during nesting, it is important to identify the causes
of egg and nestling mortality in our study population. Table
8.16 summarizes our information on this question. As is
typical of most studies, we almost never witness the loss of
an egg or nestling; therefore we must rely on circumstantial

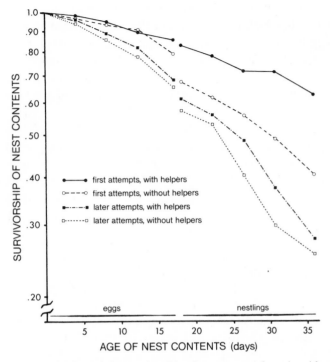

FIGURE 8.5. Cumulative survivorship of nest contents for pairs with helpers (closed) and without (open) for seasonal first nests (circles) and later attempts (squares) through the 36-day nest cycle. Seasonal first nests survive better than later attempts, and first nests attended by helpers survive best of all. Helpers do not significantly increase the survival of later attempts. Hatching failures are represented by unconnected jump between egg and nestling stage.

evidence to form our conclusions. Usually a nest that fails is found empty, but undisturbed, a day or two after last seen with eggs or healthy nestlings. We attribute such losses of entire clutches or broods to predation.

As typifies passerine birds (Ricklefs, 1969; Caccamise, 1976 and references therein), predation accounts for most losses from Florida Scrub Jay nests. Sixty-seven percent of the eggs and 85 percent of the nestlings that are lost are lost to predators. The second most common cause of egg loss is hatching failure (24% of the losses). Other causes of egg loss are negligible. A similar situation exists for nestlings, where starvation is the second most common cause of loss (13%), and all others are rare. For eggs and nestlings combined predation accounts for 74 percent of the losses.

Hatching Failure

Of 983 eggs that remained in the nest throughout the incubation period, 873 (89%) hatched. Analysis of subsets of the total sample produces a puzzling result: the probability of hatching is *significantly higher* for eggs tended by a female with helpers than by a female without helpers (table 8.17). Gross examination of the contents of some eggs that failed to hatch suggests infertility is not an explanation. For pairs with helpers embryos were seen in 9 of 20 eggs. For pairs without helpers embryos were seen in 9 of 22 eggs.

Although we do not know why hatching failure is higher for pairs without helpers, inattentiveness by the incubating female is a possibility, as this often is perpetrated by interfamilial interactions. Females readily leave their nests in order to take part in boundary disputes (Woolfenden and Fitzpatrick, 1977). It may be that breeding females in families with helpers leave their eggs less often than those without helpers. Perhaps neighbors trespass less frequently into the territories of large families. Though statistically significant and therefore a contributing factor to egg loss, hatching failure accounts for a decrease of only 8 percent from the original sample of eggs. Almost three times as many eggs are lost through predation.

Starvation

We list as starvation all losses of nestlings known to have been substantially underweight and all losses of single nest-

TABLE 8.16. Causes of Florida Scrub Jay Egg and Nestling Mortality

	Totals	Percent of Individuals	Percent of Losses
EGGS	**1337**	100.0	
Losses caused by:			
Observer damage	3	0.2	0.6
Wind	3	0.2	0.6
Fire	3	0.2	0.6
Desertion	16	1.2	3.4
Breeder competition	2	0.1	0.4
Breeding female's death	15	1.1	3.2
Predation (part of clutch)	28	2.1	6.0
(entire clutch)	284	21.2	61.0
Total losses before hatching	−354	26.5	76.3
Losses caused by hatching failure			
(embryo seen)	18	1.3	3.9
(embryo not seen)	24	1.8	5.2
(no inspection)	68	5.1	14.7
Total losses at hatching	−110	8.2	23.7
NESTLINGS	**873**	100.0	
Losses caused by:			
Flimsy nest	1	0.1	0.3
Fire	1	0.1	0.3
Breeder competition	3	0.3	0.9
Breeding female's death	2	0.2	0.6
Starvation[a]	42	4.8	13.0
Predation (part of brood)	33	3.9	10.2
(entire brood)	242	27.7	74.7
Total losses after hatching	−324	37.1	100.0
FLEDGLINGS	**549**		

[a] Includes all losses of substantially underweight nestlings and all losses of one nestling from continuing, large broods.

lings from continuing, large broods. No doubt this procedure overemphasizes starvation because such losses include young that die of injury, parasitism, genetic defects, and partial-brood predation. It also includes losses of single nestlings from large broods that occur because the nest is too crowded. Regardless, the result is a maximum loss of

TABLE 8.17. Hatching Failure Rates in Florida Scrub Jays

	Novice ♀♀		Experienced ♀♀		Totals		
	Without helpers	With helpers	Without helpers	With helpers	Without helpers	With helpers	Pooled
No. eggs that could hatch	78	57	324	524	402	583	983
No. eggs that did hatch	67	53	280	473	347	526	873
Hatching failure rate	.143	.070	.137	.097	.138	.095	.113
Chi-square	1.74(n.s.)		2.91(n.s.)		4.25 ($p<.05$)		

under 5 percent of the nestlings that hatch. This figure is similar to an earlier estimate (Woolfenden, 1978) and certainly is not large. Starvation accounts for a maximum of only 13 percent of all nestling losses.

Intuitively, the relationship between the food needed by a brood of dependent young and the food provided by the feeders would seem important to reproductive success, as has been demonstrated for many bird species (Lack, 1954, 1966). The idea seems especially appropriate for cooperative breeders such as Florida Scrub Jays where the ratio of feeders to dependent young varies from 3:1 (a feeder ratio of 3.0) down to 1:2 (a feeder ratio of .50). In order to examine the importance of starvation in Florida Scrub Jays, we adopt here an operating definition of starvation even more liberal than that used for table 8.16 by assuming that *all* nestlings lost from eventually successful broods died through starvation (table 8.18). The distribution of the resulting 53 losses between breeding pairs with helpers (5.5%) and without helpers (7.0%) is not significantly different. In contrast, with all disappearances of *entire broods* tallied as predation losses, the distribution of the results strongly supports the conclusion that pairs with helpers lose fewer nestlings to predators (26.2%) than do pairs without helpers (38.6%; $\chi^2 = 14.99$, $p < .001$).

We suspect that those losses actually caused by starvation usually do not result from delivery of insufficient food, but instead from competition for food among brood mates. Hatching is not highly synchronized in Florida Scrub Jays (Woolfenden, 1974), perhaps because incubation sometimes begins before the fourth egg of a clutch is laid. Whatever the causes, some young hatch out a day or two after their brood mates, and it appears to us that a few of these individuals lose out in competition for the food delivered, and eventually starve.

TABLE 8.18. Estimated Maximum Starvation Rates and Minimum Predation Rates for Florida Scrub Jay Nestlings

| | Total Nestlings | Nestlings Fledged | Nestlings Lost [a] | | Maximum Starvation Rates | | Minimum Predation Rates | |
			Starved [a]	Whole broods	% individ.	% losses	% individ.	% losses
Without helpers	342	186	24	132	.070	.154	.386	.846
With helpers	531	363	29[b]	139*	.055	.173	.262	.827
All pairs	873	549	53	271	.061	.164		

[a] Includes *all* nestlings lost from successful nests.
[b] Difference between pairs without and with helpers not significant ($\chi^2 = .883$).
*Difference between pairs without and with helpers highly significant ($\chi^2 = 14.989$).

Nest Predation

Having identified predation as the cause of the vast majority of egg and nestling losses, we focus briefly on the seasonal distribution of predation. As mentioned previously, Florida Scrub Jays have a short, synchronized breeding season (figure 8.1 and Woolfenden, 1974), and second and later nests of the season fail more often than first nests (figure 8.5). From these observations an increased rate of failure as the breeding season progresses is suspected. Figure 8.6, which graphs certain of the data listed in table 8.19, shows this suspicion to be true. Indeed, the probability of fledging young as the breeding season progresses decreases dramatically (figure 8.7). It may be that when the probability of fledging young falls to about 30 percent, as is the situation by June, it is better for the jays to "wait until next year." This strategy seems especially advantageous if nesting season activities decrease breeder survival, which our data suggest (see chapter nine).

8.5. IS THERE AN OPTIMAL CLUTCH SIZE?

Florida Scrub Jays lay clutches varying from 2 to 5 eggs. Nearly all clutches are either 3 or 4 (figure 8.8, top). Many factors influence the size of any single clutch in our population of jays. These include age and experience of the female (table 8.3), time in the nesting season (Woolfenden, 1974), possibly food availability during the prenesting season (see section 8.6), and apparently some genetic polymorphism within the population (unpublished data). A complete analysis of individual variation in clutch size is beyond our present scope. However, general theories of clutch size agree that some optimal number of eggs in a clutch should exist, a number that will produce the maxi-

TABLE 8.19. Two Methods for Calculating Predation Rates and Expected Nest Success through the Breeding Season

Nesting Week	Daily Predation Rates					Observed Success Ratios		
	Total nest days[a]	No. nests lost	Daily predation rate	x̄ daily predation, 36 days[b]	Expected nest success[c]	No. nests initiated	No. eventual successes	Expected nest success
Mar 1–8	34	0	.0000	.0058	.81	3	2 ⎫	.69
9–16	242	0	.0000	.0081	.75	36	25 ⎬	.59
17–24	866	7	.0081	.0105	.68	88	52	.73
25–31	1154	11	.0095	.0121	.65	56	41	.50
Apr 1–8	1568	18	.0115	.0155	.57	58	29	.40
9–15	1602	18	.0112	.0182	.52	35	14	.52
16–23	1648	20	.0121	.0213	.46	31	16	.32
24–30	1157	19	.0164	.0227	.44	25	8	.29
May 1–8	1026	27	.0263	.0273	.37	31	9	.24
9–15	677	17	.0251	.0292	.34	17	4	.44
16–23	720	19	.0264	.0318	.31	23	12 ⎫	
24–31	516	10	.0194	.0332	.30	4	0 ⎬	
Jun 1–7	329	13	.0395	.0378	.25	—	—	—
8–15	198	7	.0354	—	—	—	—	—
16–23	104	4	.0385	—	—	—	—	—
24–30	2	0	—	—	—	—	—	—

[a] Combined nest days from 10 breeding seasons, 1970–1979.
[b] Mean daily predation rate calculated as arithmetic mean of 5 weekly mean rates, beginning with week of nest initiation.
[c] Expected nest success = (daily survival rate)[36]; daily survival rate = 1 − (x̄ daily predation rate).

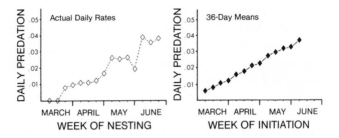

FIGURE 8.6. Two plots of daily rates of nest predation through the breeding season. Graph at left plots the weekly averages of actual daily rates of nest loss, where rate is calculated from the number of nests lost in that week and the number of nest days in the weekly sample. Graph at right shows the mean daily predation rate to be expected over the 36-day period following initiation of incubation.

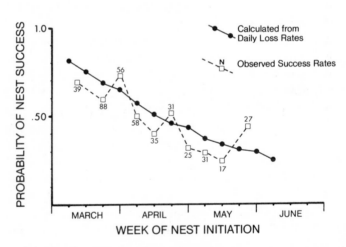

FIGURE 8.7. Probability that a nest, initiated at a given date, will fledge young. Plotted are raw data showing observed success rates (open squares) and calculated probabilities (closed circles) based on daily loss rates shown in figure 8.6. Nests are not initiated after the last week of May, and final sample (N = 27) pools last two weeks of that month.

mum number of surviving offspring (Lack, 1954, 1966; Cody, 1966; Ricklefs, 1968; review by Hutchinson, 1978). We have seen that fledgling survival at least to age 1 year is independent of fledgling brood size (table 8.12). Therefore, in the Florida Scrub Jay, we can make a preliminary search for an optimal clutch size by comparing the average fledgling productivity among clutches varying from 2 to 5.

Figure 8.8 (top) shows the frequency distribution of clutch sizes in our population, and (bottom) the average fledgling production from each clutch size class. Among nests that successfully produce at least one young, productivity is directly proportional to clutch size (solid line). However, a lower proportion of successful nests among 5-egg clutches (dotted line, not significant) results in exactly equal average numbers of fledglings from 4- and 5-egg clutches when *all* clutches are compared (dashed line). If the trend is real, it perhaps explains why 5-egg clutches are rare. The extra egg adds nothing to *average* expected fledgling production, and even somehow may reduce the probability of fledging any young at all. (Note, however, that the curve for *successful* nests suggests why clutches of 5 might be favored under particularly favorable circumstances.)

The question of why the modal clutch size is 3, even though clutches of 4 show equivalent success rate and produce significantly more fledglings, remains unanswered. Table 8.20 shows that clutches of 4 produced more fledglings in eight of the ten years in our sample. In one year (1972) the clutches performed equally, and in only one year (1978) did clutches of 3 outproduce clutches of 4. Because probability of failure is roughly equal between the two clutch sizes (figure 8.8), these preliminary analyses suggest that 4-egg clutches should be favored in our population. Testing the effects of breeder experience, age and stability of pair bonds, social environment, and a host of population genetic

217

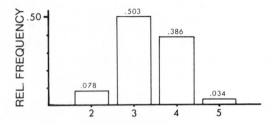

FIGURE 8.8. Frequency distribution of clutch sizes (top) and average fledgling production and nest success for each clutch size (bottom). Nest success (dotted line), the proportion of nests fledging at least one young, is roughly equal for all clutch sizes. Among successful nests, fledgling production is directly proportional to clutch size (solid line). When all nests are included (dashed line), fledgling production increases from clutches of 2 to clutches of 4, but is about equal for clutches of 4 and 5. Sample sizes and standard errors are shown for both comparisons; standard deviations are also shown for the sample containing only successful nests.

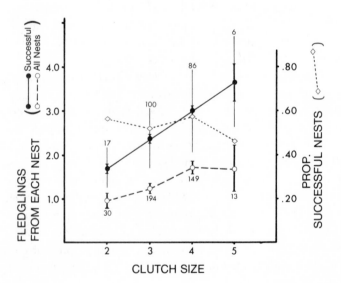

TABLE 8.20. Average Fledgling Production per Nest Attempt as Related to Individual Clutch Size, 1970–1979

Clutch Size	Breeding Season										Annual Mean
	'70	'71	'72	'73	'74	'75	'76	'77	'78	'79	
2	1.3(3)	2.0(1)	1.0(3)	0.0(2)	0.8(4)	—	1.7(3)	0.6(5)	1.0(5)	1.0(4)	1.04
3	2.0(6)	1.2(14)	0.5(27)	1.0(21)	0.7(12)	1.5(15)	1.0(26)	1.7(17)	1.7(27)	1.7(25)	1.31
4	2.8(9)	1.8(12)	0.5(10)	1.6(11)	1.6(22)	2.1(21)	1.3(15)	2.7(11)	1.2(18)	2.6(16)	1.82
5	—	—	3.0(1)	0.0(2)	1.0(3)	1.0(4)	4.0(1)	—	—	3.0(2)	2.00

NOTE: Excludes 1969 data included in figure 8.8. Sample sizes for each clutch size shown in parentheses.

properties upon this question awaits the larger sample sizes provided by future years' study.

8.6. ANNUAL VARIATION IN REPRODUCTIVE SUCCESS

Throughout this chapter we have measured and compared Florida Scrub Jay reproductive success by calculating the *average number of fledglings produced per pair* for various samples of pairs. We showed in figure 8.2 that although pairs with helpers consistently fare better than those without in any given year, the absolute level of productivity varies widely from year to year. The data in table 8.2 produce coefficients of variation in annual fledgling production per pair of 28.4 percent in the population as a whole, 42.8 percent among pairs without helpers, and 25.3 percent among pairs with helpers. Identifying the cause of this annual variation is of considerable importance in our effort to describe the biology of the population and perhaps to further isolate the exact role of helpers in affecting reproductive output.

Annual variability in fledgling production could proximately arise from any combination of variations in four principal factors: (1) clutch size, (2) hatching failure, (3) loss of nestlings through starvation, or (4) loss of eggs or nestlings through predation. Accidental losses from wind, fire, poor nest construction, and the like are too infrequent to be responsible for gross annual differences (see table 8.16). Ultimately, any one of the above four factors could, in turn, vary with annual differences in environmental variables such as moisture, food availability, and vegetation cover characteristics. Below, we examine variations in each of the four factors.

Clutch Size

In our population, mean clutch sizes for the first clutches of the season for all pairs varied over 10 years from 3.07 to 3.88 (annual $\bar{x}$ = 3.37, CV = 7%). Annual variation is reduced somewhat considering all nest attempts and not just first clutches (range = 3.21 to 3.74; $\bar{x}$ = 3.38, CV = 5%). This variation could help cause fledgling production to vary, because the number of fledglings produced from an individual (successful) nest strongly depends on how many eggs were laid to begin with (figure 8.8). Hence, during years when mean clutch size is high or low, mean fledgling production might be expected to vary the same way.

The importance of clutch size variation is further implied by the evidence that it is strongly influenced by the previous summer's rainfall (figure 8.9a), presumably by causing variation in insect food abundance before egg-laying (untested). It is surprising, therefore, that annual fledgling production actually shows little dependence upon mean clutch size (figure 8.9b). The slight positive trend is not significant, and were it not for a single, especially high data point (1975) the trend would be invisible. We must search elsewhere for a primary source of variation in annual productivity.

Hatching Success and Nestling Starvation

Annual variations in reproductive success could result from year-to-year variations in hatching success rate or in nestling starvation rate. The first of these can be tested directly by plotting annual reproductive outputs against the respective hatching success rates for each year (figure 8.10a). No correlation emerges beyond a slight and insignificant positive trend. The starvation question is addressed by plot-

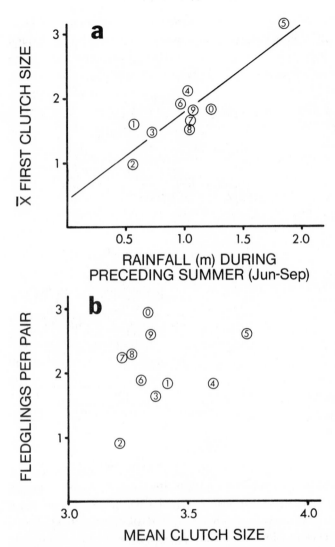

FIGURE 8.9. Correlation ($r^2 = .80$) between rainfall and mean first clutch size (8.9a), and absence of a correlation between mean clutch size and fledgling production (8.9b). The circled numbers, 0–9, represent the years 1970–1979.

ting the same annual reproductive measure against the survival proportions of nestlings among all broods that fledged at least one young (figure 8.10b). This conservative approach assumes that all losses from continuing broods resulted from starvation. Again, despite a slightly positive trend, no relationship emerges. Thus, neither hatching failure nor starvation alone accounts for much annual variation in reproductive success.

To test whether the above three variables *together* can account for variation in fledgling productivity, we subjected the data to a multiple regression analysis using clutch size, hatching success, and nestling survival in successful nests as three independent variables. Once again, we failed to find any significant relationship (correlation coefficient = .44, r^2 = .19, F ratio = 1.56).

It is possible that with additional years' data some significance to the above-mentioned trends may be found. However, 10 years' data demonstrate that no clear correlation is likely to emerge that will relate annual fledgling productivity to mean clutch size in the population or to nonpredator-related egg and nestling mortality (figures 8.9, 8.10). As shown below, the overriding cause of annual variation in reproduction apparently lies in variable predation rates.

Nest Predation

Our best index for annual incidence of nest predation is the proportion of nests that survive to fledgling. Eggs do not starve and are virtually never abandoned; hence any nests with eggs in which no eggs survived to hatching almost certainly was preyed upon. A few nestlings do starve (table 8.18), but starvation does not affect *entire* broods except in rare and atypical cases. Therefore, we classify the simultaneous disappearance of an entire brood as predation.

223

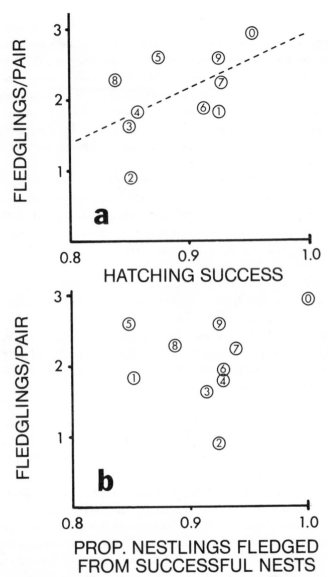

FIGURE 8.10. Relationships between annual fledgling production per pair and both hatching success (8.10a) and proportion of nestlings fledged from successful nests (8.10b). The circled numbers, 0–9, represent the years 1970–1979. Dashed line shows positive, but not significant, correlation.

Annual variation in nest survival has been extensive during the decade, ranging from .267 in the worst year (1972) to .778 during the best year (1970). We further investigate the importance of predation by plotting annual fledgling production per pair against our figures on nest survival (figure 8.11a). The overwhelming importance of predation rates shows clearly in the nearly linear relationship between these measures.

In figure 8.11b we plot fledgling production against a slightly different measure of nest survivorship, namely, the average *life-span* of all nests attempted during each year. The interpretation of this measure is made easier by examining figure 8.12, which tracks the progressive destruction of active nests by predators over the 36-day potential nest life-span, for 4 separate years for which we have extensive nest survival data, and for the pooled 10-year sample. During the poorest years (e.g., 1972, 1973) nest destruction occurred more frequently and at earlier stages than during the best years (e.g., 1975, 1979). This pattern results in quite different average life-spans during the various years (figure 8.11b), and accounts for virtually all of our measured variation in per-pair fledgling production.

A final indication of the importance of nest predation in determining fledgling productivity is contained in figure 8.13, where nestling and fledgling *brood sizes* are shown to be positively correlated with the proportion of nests that survived to fledgling. If we accept the latter measure to be an independent, inverse index of annual predation rate (see above), figure 8.13 suggests that even among nests that avoid complete predation, the size of the resulting broods is largely dependent upon the same factor. These plots indicate that clutch and brood reduction by predators accounts for much of our observed annual variation in brood size.

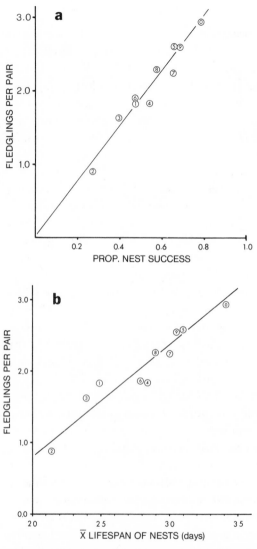

FIGURE 8.11. Correlations between annual fledgling production per pair and both proportion of nests fledging at least one young (8.11a; $r^2 = .95$) and the average life span of all nests of a season (8.11b; $r^2 = .91$). The circled numbers, 0–9, represent the years 1970–1979. Linear regression lines shown. Nearly perfect correlations indicate overriding importance of nest predation to annual variations in reproductive success.

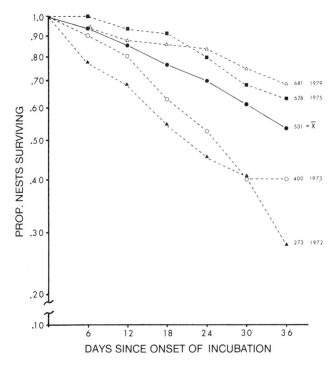

FIGURE 8.12. Average nest survivorship throughout the nest cycle during the two poorest years (1972, 1973) and the two best years (1975, 1979), and for all 10 years combined (solid line).

The above data show that predation rate during the nesting season determines the average number of fledglings produced per pair by affecting overall nest success *and* the number of fledglings produced from successful nests. The reasons for wide annual fluctuations in predation rate are unclear at present. One interesting environmental correlate warrants mention, however. Figure 8.14 shows that the proportion of nests producing at least one fledgling significantly correlates with rainfall during the 10 months preceding the nesting season. Rainfall could influence nest

227

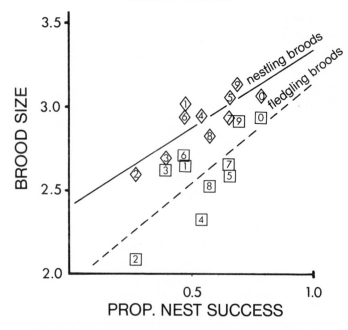

FIGURE 8.13. Correlations between proportion of nests fledging at least one young and brood sizes of both nestlings (diamonds; $r^2 =$ .69) and fledglings (squares; $r^2 =$ .53). Enclosed numbers, 0–9, represent the years 1970–1979. Linear regression lines shown through each set of points.

predation by affecting (1) the density or activity of nest predators, (2) the availability of alternate food sources for the predators, (3) the time budgets and nest vigilance of the jays, or (4) the vegetation cover surrounding nests. At present, we can present the correlation only as a basis for future research in this area.

8.7. FECUNDITY AND AGE OF BREEDER

Earlier we showed that the production of eggs (table 8.3) and fledglings (table 8.4) increases slightly from novice,

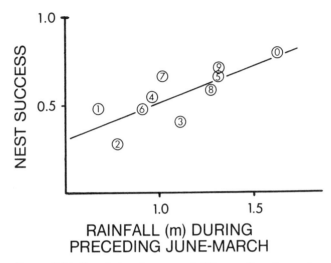

FIGURE 8.14. Correlations between rainfall preceding the nesting season and proportion of nests fledging at least one young (r^2 = .61). Circled numbers, 0–9, represent the years 1970–1979. Linear regression line shown.

through mixed, to experienced pairs. Within all three categories, pairs with helpers produced no more eggs, but substantially more fledglings, than did pairs without helpers. Of great importance to demography is knowing more precisely how much fecundity changes with age of the breeders. Nearly all existing comparisons in the avian reproduction literature are made between novice (female) breeders and all of the experienced (female) breeders combined (e.g., Coulson, 1966; DeSteven, 1978; review by Ricklefs, 1973; Harvey et al., 1979 [who include males]). Our data allow us to examine the question of age-specific fecundity for several consecutive years of breeding, instead of the less refined two-way comparison.

Table 8.21 summarizes data on egg and fledgling production by females with from 0 to 9 or more years of breeding experience. Because clutch size was shown not to vary significantly with presence or absence of helpers (table 8.3), we combine these data in table 8.21. First clutches are extracted from the overall means and shown separately, because these are known to differ from the seasonal average during some years, and first clutches account for most of the young produced (see section 8.5). Because helpers are known to increase fledgling production, we separate females without helpers from females with helpers in presenting the fledgling data.

Overall mean fecundity, measured both in clutch size and in fledgling production, is graphed against breeder experience in figure 8.15. This graph pools all females of each level of experience regardless of the presence or absence of helpers. The standard error is shown for each mean value. This summary graph clearly displays a trend toward increased fecundity with increased breeding experience, both in clutch size and in fledgling production. From the point of view of individual females over an average lifetime, net reproductive potential increases with age at least over the first 4 years of breeding. Clutch size may level off after year 5, and even may decline slightly as females age, but our samples to date prohibit conclusive analyses of the older age classes.

As in previous analyses, variance in fledgling production is extremely high, primarily because a relatively large proportion of nest attempts fail entirely. Combined with marginal sample sizes, this variance tends to swamp the apparent jump in fledgling production between years 2–3 and years 4 through 6 of breeding (figure 8.15). However, it is apparent that even in years 2–3 of breeding, average fledgling production is not at its maximum potential. This pattern holds whether or not helpers are present, even though a

TABLE 8.21. Fecundity in Relation to Duration as a Breeder among Female Florida Scrub Jays

	Breeding Year[a]								
	1	2	3	4	5	6	7	≥8[b]	≥9[b]
Clutch size (all)									
Mean	3.06	3.25	3.35	3.47	3.57	3.36	3.25	3.31	3.25
S.D.	.52	.69	.81	.70	.51	.81	.89	.74	.75
	48	36	20	19	14	11	8	26	12
First clutch only									
Mean	3.06	3.19	3.39	3.42	3.70	3.13	3.17	3.39	3.33
S.D.	.56	.68	.96	.79	.48	.83	.98	.78	.71
	33	27	13	12	10	8	6	18	9
Fledglings (females without helpers)									
Mean	1.38	1.57	2.00	1.80	1.67	2.33	2.50	1.67	—
S.D.	1.53	1.22	1.60	1.79	1.53	1.15	—	1.53	—
	24	14	8	5	3	3	2	3	—
Fledglings (females with helpers)									
Mean	2.21	1.87	1.38	2.56	2.43	2.00	2.20	3.25	3.20
S.D.	1.85	1.06	1.19	1.13	1.13	2.19	1.48	1.04	1.99
	14	15	8	9	7	6	5	8	10

[a] Table begins with first year of breeding, which almost always corresponds to age 2 or 3 years among females (see table 4.2). Years 1 through 7 include known-age females only.

[b] Samples of breeders at least in their 8th or their 9th years of breeding include some unknown-age females banded as breeders early in the study and surviving this long.

greater increase is shown among the sample with helpers (table 8.21).

8.8. HYPOTHESES FOR HELPERS BENEFITING FROM HELPING

Two obvious potential benefits to the individuals who forgo breeding and remain home as helpers are (1) increased reproductive success following helping and (2) gains

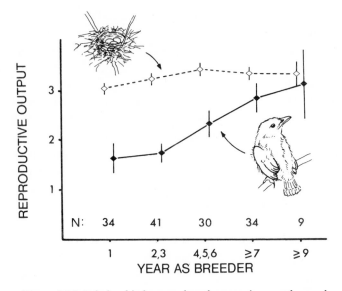

FIGURE 8.15. Relationship between breeder experience and annual reproductive output, as measured by clutch size (open diamonds) and mean number of fledglings (closed diamonds). The vertical lines show one standard error on either side of the means.

in inclusive fitness obtained indirectly by raising sibs or half-sibs. Our data are sufficient to offer insight into both of these questions.

Does Helping Increase Breeding Success of Helpers?

Jays that remain at home as "apprentices" instead of attempting breeding at age 1 year may gain through later increases in breeding success or through increased longevity measured over the lifetime of the individual. Unfortunately, we have no true control for testing this possibility, because virtually all young jays help at least one year before becoming breeders. Furthermore, our sample of lifetime reproductive histories of known-age jays still is far too small to permit analysis of this critical measure. We can, however,

compare annual rates of reproduction and survival between jays that helped only one year and those that remained two or more years (table 8.22, figures 8.16a and
8.16b).

It has long been suggested that experience as a helper
might improve later reproduction, presumably through the
effects of "practice" at breeding (Skutch, 1961; Lack, 1968;
Brown, 1978a). To our knowledge, table 8.22 is the first
numerical indication in any bird that such an improvement
actually might occur. Among males, breeding success shows
the familiar increase from the first year of breeding up
through the fourth year and beyond (reading down the
columns in table 8.22; cf. table 8.21). However, males who
helped only one year increase from averages of 1.4 to 2.6
fledglings per year, whereas males who helped three or
more years average 2.0 fledglings *their first year,* and show
an increase to an impressive average of 2.8 by their fourth
year. Proportional nest success shows a parallel trend. The
results are similar for female helpers, among whom we
could compare only two samples of helper experience.

We emphasize the preliminary nature of these results.
Only one of the numerous comparisons we made shows a
statistically significant difference related to helping experience (table 8.22, footnote a). Furthermore, the seeming
equivalence between 1 and 2 years' helping experience
among male helpers is not consistent with the hypothetical
trend. Nevertheless, table 8.22 suggests that a real component of beneficial experience is involved in helping. The
fact that such exists even between relatively old helpers
leads us to suggest that the benefits might be even *more*
important at age 1 year. Helpers, especially young ones,
gain valuable experience at such necessities as foraging and
food-caching, predator avoidance, territory defense, and
nesting. We suspect that this experience easily could nullify

233

TABLE 8.22. Reproductive Success of Breeders in Relation to Time Previously Sp
Helpers

| Year of Breeding | Years as Helper (Males)[a] | | | Years as Helper (Fe |
	1	2	≥3	1	
First	1.43[b] ± 1.62	1.42 ± 1.44	2.00 ± 1.35	1.46 ± 1.47	1.69
N[c]	23	12	12	24	
Prop. success[d]	.48	.58	.83	.54	.
Second-third	2.04 ± 1.59	1.57 ± 1.34	2.42 ± 1.38	1.69 ± 1.18	1.75
N	25	14	12	32	
Prop. success	.76	.64	.83	.75	.
Fourth	2.56 ± 1.94	2.50 ± 3.21	2.83 ± 1.17	2.24 ± 1.46	2.50
N	25	6	6	34	
Prop. success	.80	.50	1.00	.79	.

[a] Proportional success by males with 3 or more years helping experience (.87, N
significantly higher than that of males with 1 or 2 years helping experience (.66, N
$\chi^2 = 4.91$, $p < .05$).
[b] Entries show mean number of fledglings produced in each season ± 1 standard de
[c] Sample sizes represent number of breeding seasons monitored for each categor
[d] Proportional success = fraction of breeding seasons in which at least 1 fledgl
produced.

any "sacrifice" they are making to their fitness by not breeding. More data are required before we are willing to offer this suggestion in any stronger form.

In startling contrast to the results just described, the jays with only one year of helping experience before becoming breeders show dramatically *higher* survival rates than the jays that help for two or more years before breeding. Higher survival is suggested by two analyses. First we plot *survival as breeders* for jays that helped one year (closed circles) and for more than one year (open circles), regardless of the actual age of the jays (figure 8.16a). The difference between the two rates is significant, as measured by the linear regressions run on the two samples ($t = -2.83$, $p < .05$). As estimated by these regressions, jays who helped only one

234

year show about .90 annual survivorship as breeders, while those helping longer before breeding show about .77. However, this analysis is valid only if survival is age independent once jays become breeders. As we have some evidence to the contrary (see chapter nine), we attempt another, more complicated analysis that measures *survival relative to real age* for jays that helped for only one year (closed circles) and for more than one year (open circles) (figure 8.16b). Naturally, death prior to first breeding prevents an individual from entering the sample. As prebreeding deaths should affect potential late first breeders more than potential early first breeders, the survival rates plotted for late breeders in figure 8.16b are maximum estimates. Therefore the results we obtain are conservative. Again the difference between the two rates is significant and substantial (.90 versus .72 annual survival; $t = 7.28$, $p < .001$).

These results strongly suggest that no long-term *survival* value is obtained from additional years at home after the first one. Further, the longevity data offer our first indication that jays who *do* remain at home for two or more years actually may include individuals that are in some way either less healthy or for some reason less capable than those that help for one year only. Further analysis of this intriguing pattern awaits larger sample sizes of jays helping more than the usual one year.

Is Raising Sibs as Good as Raising Offspring?

In his review of cooperative breeding in birds, Emien (1978) used our 1970 to 1977 jay data to compare the contributions to an individual's inclusive fitness gained from helping raise siblings with the contribution that could be expected if the individual "immediately" became a breeder. As is mentioned, a true control is not available in our population because young birds virtually never attempt breed-

235

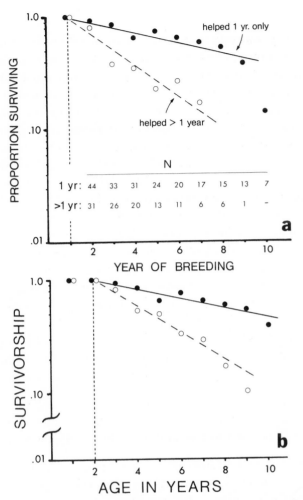

FIGURE 8.16. Survivorship as breeders (8.16a) and simply as individuals, breeding or not (8.16b), among jays that helped only one year (closed circles) versus two or more years (open circles). Original samples (N) consist of jays that eventually did become breeders for at least one year, hence year 1 of breeding (8.16a) and year 2 of life (8.16b) by definition show perfect survival. Each survivorship curve shows approximately age-independent mortality thereafter, but the two differ significantly from each other in both graphs. Sample sizes refer to the number of known-age breeding individuals that could have been observed surviving x number of years after becoming breeders.

ing at age one year (chapter four). By using novice breeders as the control, however, an approximation of Emlen's N_0 (number of young produced by a novice individual during its initial breeding attempt) was obtained for his comparison. In reality, this approximation represents a maximum estimate of the true N_0, because the figures were not corrected for any reduced probability of successful dispersal by one-year-old jays. Emlen's estimate of inclusive fitness gains through helping also was high, because on the average helpers are not related to the nestlings they help raise by a factor of .50 (r_h of Emlen). Parental deaths and replacements by stepparents result in a lowering of this average degree of relatedness to .43 (see chapter ten).

Table 8.23 is presented in the format of Emlen (1978, and his table 9.3), and shows our most recent calculations of the pertinent parameters for his analysis. The results are similar and the conclusion unchanged. *Young jays—if they can become breeders—will add significantly more to their own inclusive fitness by breeding rather than helping.* More detailed analysis of this important, but complicated, question is presented in chapter ten.

8.9. PER CAPITA REPRODUCTIVE SUCCESS

Several investigators studying cooperative breeding in birds (notably Koenig, 1981; Koenig and Pitelka, 1981) stress the importance of measuring reproductive success of different-sized groups on a per capita basis, rather than per breeding unit, or per group, as we have done above. Koenig (1981) argues that this constitutes a measure of how much each individual bird gains by dwelling in a group, which then can be compared to its potential gains were it to breed as a member of a simple pair. Although we dis-

TABLE 8.23. Effects of Experience and Presence or Absence of Helpers on Reproductive Success, with Calculation of Fitness Gains to Helpers

	Fledglings per Year	N (Pairs)	Proportion Successful	Category
N_2	2.38	117	.80	Pairs where both members have prior breeding experience; helpers present.
N_1	1.80	64	.73	Pairs where both members have prior breeding experience; no helpers present.
$\overline{H}$	1.78			Average number of helpers present in groups with helpers.
N_0	1.76	87	.66	Pairs where at least one member is a novice breeder.
N_0^1	1.31	52	.58	Subsample of N_0 where no helpers present.
N_0^2	1.24	17	.47	Subsample of N_0 where both breeders are novices and no helpers present.

$r_h = .43; r_p = .50$

*Case 1: $\dfrac{N_2 - N_1}{\overline{H}} \times r_h \quad < N_0 \times r_p; \quad .14 < .88$
$(.25)$

Case 2: $\quad "\quad < N_0^1 \times r_p; \quad .14 < .66$
$(.25)$

Case 3: $\quad "\quad < N_0^2 \times r_p; \quad .14 < .62$
$(.25)$

Note: Format follows Emlen, 1978, his table 9.3.
*Fitness gain calculated for one helper by dividing the mean number of helpers (unadjusted figure in parentheses shows fitness gain for a lone helper).

agree with this assertion and think that Koenig's (1981) arguments are flawed (see below), we present these per capita reproductive data for comparative purposes in table 8.24.

Per capita reproductive success in the Florida Scrub Jay is equal between simple pairs and pairs with one helper. A clear and highly significant decline in this measure occurs as family size increases beyond three (table 8.24, right-hand column). This result is consistent with those summarized by Koenig (1981) for a variety of cooperative breeders. Koenig uses this pattern as evidence that group sizes greater than two are in some sense suboptimal. However, we see no biologically relevant reason for computing reproductive success in this manner, at least in our system, for two primary reasons.

1. We have no evidence, after 13 years of study, that any jays besides the pair contribute eggs or sperm to the contents of a nest. It is true that most helpers (64%) are on average equally related to the siblings they help raise and to any of their own young they might raise ($r = .50$). However, this is by no means always the case (see table 5.3). The fact that helpers continue to help despite frequent changes in their relatedness to nestlings (figure 5.2) means that the nestlings are on the average considerably more valuable to the breeding pair than they are to the helpers (see table 8.13), from the strictly genetic point of view. Simple per capita calculations that ignore this inequality overestimate helper gains relative to breeder gains when inclusive fitnesses of the two classes are calculated. We question the assertions of Koenig and Pitelka (1981) that the breeding structure within the group bears nothing on the validity of such calculations.

In contrast to Florida Scrub Jays, some Acorn Woodpecker nests sometimes contain eggs from two or more

TABLE 8.24. Mean Annual Per Capita Reproductive Output within Different-sized Florida Scrub Jay Groups, 1970–1979

Number of Helpers	Breeding Season										Average No. Fledglings per Capita	
	'70	'71	'72	'73	'74	'75	'76	'77	'78	'79	Individual means	Annual means[a]
Zero	1.15	0.50	0.16	0.68	0.73	1.16	0.46	0.93	1.00	1.04	0.79	0.78
S.D.	0.56	0.71	0.43	0.75	0.64	0.60	0.61	0.61	0.79	0.72	0.63	0.33
N	7	4	13	14	13	16	11	13	14	12	117	10
One	1.11	0.83	0.47	0.58	0.85	1.07	0.78	1.00	0.76	0.88	0.81	0.83
S.D.	0.19	0.53	0.45	0.46	0.58	0.64	0.69	0.33	0.50	0.63	0.51	0.20
N	3	8	5	8	9	5	6	3	11	16	74	10
Two	1.00	0.33	0.33	0.75	0.38	0.56	0.60	0.61	0.75	0.85	0.57	0.62
S.D.	—	0.26	0.44	0.25	0.25	0.43	0.34	0.33	0	0.55	0.34	0.23
N	2	6	6	3	4	4	5	9	3	5	47	10
Three	0.70	0.30	0.30	—	0.00	0.80	0.40	—	0.60	0.70	0.48	0.48
S.D.	—	—	0.26	—	—	—	0.40	—	0.72	—	0.39	0.27
N	2	2	4	0	1	2	3	0	3	2	19	8

TABLE 8.24 (continued)

| Number of Helpers | Breeding Season | | | | | | | | | | Average No. Fledglings per Capita | |
	'70	'71	'72	'73	'74	'75	'76	'77	'78	'79	Individual means	Annual means
Four	—	0.50	—	0.00	—	—	0.47	0.50	0.33	—	0.41	0.36
S.D.		—	—	—	—	—	0.14	—	—	—	0.19	0.21
N	0	1	0	1	0	0	5	1	1	0	9	5
Five	—	0.29	—	—	—	—	—	—	—	0.29	0.29	0.29
S.D.		—	—	—	—	—	—	—	—	—	—	—
N	0	1	0	0	0	0	0	0	0	1	2	2
Six	—	—	—	—	0.38	—	—	—	—	—	0.38	0.38
S.D.		—	—	—	0.38	—	—	—	—	—	—	—
N	0	0	0	0	1	0	0	0	0	0	1	1

NOTE: Per capita values determined by dividing per-pair measures, such as those summarized in table 8.8 by group size. Sample sizes (N) show number of groups in each category.

[a] Right-hand column shows averages of the annual mean values and their standard deviations; in this column, N represents number of years in sample.

females (Koenig and Pitelka, 1979; Koenig et al., 1983), and several male Acorn Woodpeckers may breed within one group (Stacey, 1979b; Koenig, 1981). Therefore, the genetic relatedness between individual group members and the young produced by the group must vary widely even within the Acorn Woodpecker. Certainly the fitness value of a given fledgling must vary even more *between* species, the primary source of this variation being differences in the structure of the breeding unit. Therefore, we conclude that if per capita reproductive measures have different fitness values between species, and even between groups *within* a species, this is not a valid currency by which to compare fitness.

2. Any explanation of group breeding must examine and compare inclusive fitness of alternative strategies (e.g., Hamilton, 1964; Alexander, 1974; West Eberhard, 1975; Emlen, 1978; Brown and Brown, 1981b). However, fitness cannot be determined simply by counting offspring or "offspring equivalents" within the expected lifetimes of individuals. Inclusive fitness is a *relative* measure of how successful a given strategy (and its associated genome) is at spreading through a population. Factors such as delayed breeding and age-specific changes in survival probabilities are intimately interwoven with any measure of fitness (e.g., Leslie, 1966). Therefore, per capita measures of reproductive success—even if they *are* comparable between breeders and helpers (they are not, as argued above)—do not tell the whole story of inclusive fitness. Koenig (1981, figure 1) must assume that they do for tests of his model to hold. However, he tests a model that requires figures for "lifetime reproductive success" (op. cit., p.1) with data on single-season reproductive output, calculated per capita. In our view these single-season measures alone are insufficient for building any general theory about the evolution of coop-

erative breeding. Further elaboration of these points appears in our conclusions to follow (section 8.10) and in chapter ten.

8.10. CONCLUSIONS

Assessment of the degree to which helpers contribute to offspring and breeder fitness is fundamental to any study of cooperative breeding. Although the effects of helpers rarely will be restricted to reproduction, virtually all available literature focuses on this aspect of their cooperation. This approach is most common because it is relatively easy to measure nesting success, but difficult to measure most, if not all, of the other ways that helpers might affect breeders. In this chapter, besides summarizing overall fecundity patterns in the population, we have emphasized the influence that helpers have on fledging success. Equally important are the influences of helpers on territory size and stability (chapter six) and on fledgling and breeder survival (see chapter nine).

In many respects the impact of helpers on breeder reproduction can be measured more accurately in Florida Scrub Jays than in other cooperative breeders. The most important reason is that we are almost never in doubt regarding the parentage of the young born in our study tract. In many of the best-studied cooperative breeders (e.g., Acorn Woodpecker, Pukeko, Galapagos and Harris hawks), determining parentage of the individual offspring often is impossible. Another reason is that each year many pairs breed without helpers and often the same pairs *have* helpers some years but not others. Thus, Florida Scrub Jays have an intrinsic natural control for testing the effects of helpers on breeding. Directly measuring the effects of helpers on breeders under natural conditions is, of course, impossible for species that never breed naturally as simple pairs.

We here present our conclusions on three critical questions regarding Florida Scrub Jay helpers. Do the helpers help breeders fledge more young? When in the breeding cycle and the nesting season do the helpers help? How do the helpers help? We end this discussion with comments on the limited usefulness of annual reproductive measurements as theoretical parameters in modeling cooperative breeding strategies.

One principal result of our study is unequivocal: breeders assisted by helpers fledge more young. The annual advantage during the decade averaged .8 fledglings, or half again as many young as are fledged per year by unassisted pairs. Good and bad years for reproduction affect pairs with and without helpers similarly, and the overall relationship has held for each year in the 1970s. The advantage gained by nesting pairs with helpers does not increase through 12 months of postfledging survival, but it does persist. Pairs with helpers have .3 more yearlings, a 50 percent increase over unassisted pairs. Various other analyses demonstrate that helpers in effect provide reproductive "insurance" by increasing a pair's probability of succeeding during a given season and producing at least one helper to assist during the following season.

Because helpers usually are the offspring of breeders, the above results could hold even if helpers really provide no assistance at all. Previously successful breeders tend to have helpers while pairs without helpers include most of the novices, thereby severely biasing our comparisons. By dividing our sample into categories based on experience of the breeders (i.e., experienced, novice, and mixed pairs) we removed the bias, and the result was unchanged. Helpers do help, regardless of breeder experience.

Territory quality varies, and this factor also could produce a correlation between presence of helpers and higher reproductive success, as has been shown for other cooperative breeders (Brown and Balda, 1977; Brown and Brown, 1981b; Gaston and Perrins, 1974). We minimized the effect of variable territory quality by comparing fledging success only for those breeders that bred in essentially the same territory for two or more years, both with and without helpers. Again breeders fledged more young when they had helpers. In even more restrictive tests, using the same *individual pairs* in the same territories, the presence of helpers still correlated with the production of more fledglings, although in this comparison the difference fell short of statistical significance.

Among many other species that possess the natural control of pairs breeding both with and without helpers, a similar increase in fledgling production with helpers has been reported (e.g., Harris Hawk, Pukeko, Tasmanian Native Hen, Kookaburra, Puerto Rican Tody, Red-throated Bee-eater, Superb Blue Wren: references in Brown, 1978a; also Galapagos Mockingbird: Kinnaird and Grant, 1982). A decrease in fledgling production with helpers is reported only for the Dunnock (Birkhead, 1981). According to Zahavi (1974), larger groups of the Arabian Babbler did not fledge more young than the smaller groups. Brown (1975b) questions this conclusion. No correlation is evident either way for some species (references in Brown, 1978a).

Brown et al. (1982) report the details of the first removal experiment to corroborate the above results in another species, the Grey-crowned Babbler. Nine large breeding groups (one pair plus six to eight helpers) reduced to trios (the pair plus one helper) fledged fewer young from second and subsequent nests of the season than did 11 unaltered large groups. How these babbler helpers helped remains to be determined.

In Florida Scrub Jays, helper constituency appears not to influence fledgling production dramatically. Neither the number of helpers beyond one, nor sex of the helpers, correlates with fledgling production. Age of helper does correlate positively with fledgling production, but not significantly. As older helpers are more attentive than younger ones (Stallcup and Woolfenden, 1978), this correlation may prove significant with larger samples, and probably is biologically real even if small. An increase in fledgling production with an increase in number of helpers beyond one *has* been shown for several species (e.g., Arabian Babbler, Grey-crowned Babbler), although unsolved problems still exist in which more than one female may lay in the same nest (e.g., Acorn Woodpecker). Many species show little increase in fledgling production as group size increases (Koenig, 1981). The fact that additional, "superfluous" supernumeraries frequently care for young, even though doing so does not increase fledgling production, underscores the importance of benefits to helping behavior beyond the indirect fitness benefits of helping to produce close kin (see chapter ten).

When do helpers help with reproduction in the Florida Scrub Jay? They help eggs to reach hatching and nestlings to reach fledging, but they barely help fledglings to reach independence and do not help independent young to become yearlings. Their presence increases nesting success only early in the season. Later in the season, when most breeders are attempting second or later nestings, helpers either help less or their help is swamped by the external forces that cause most nesting failures. The existence of these complexities (also encountered by Brown et al., 1982) suggests that great caution be used when measuring nesting success with and without helpers in other species.

How do helpers help the breeders fledge more young? Analyses of how nests fail show that helpers help primarily by reducing nest predation. At least 67 percent of the eggs and 85 percent of the nestlings *that are lost* are taken by predators. Egg and nestling predation is significantly lower for pairs with helpers than pairs without helpers. How this difference arises we do not know, although we speculate that breeding females with helpers may be more attentive to their eggs because they respond to fewer territorial disturbances, and their attentiveness increases hatching success. Regardless, hatching failure is a minor source of loss. Starvation of the jay nestlings also is rare; it accounts for less than 6 percent of all nestling losses. Furthermore, we suspect that some of the nestlings that appear to starve do so because they are outcompeted by their brood mates, and not because the brood is delivered an insufficient food supply. Regardless of these conclusions, no significant relationship exists between presence or absence of helpers and assumed starvation of nestlings. The relative rarity of hatching failure and starvation directs our attention back to predator dissuasion as the way that helpers help the breeders fledge more young.

Little information has been obtained on the nest predators of Florida Scrub Jays. Certainly included are some snakes, mammals, and birds. Florida Scrub Jays cannot dissuade nocturnal predators, nor probably certain diurnal nest predators, such as some snakes, once they have found the nest. However, mobbing may alter the course of foraging of diurnal predators, including snakes and most birds. Certainly other predators such as the corvids could be driven from the nest even during predation. The mobbing of large snakes by Florida Scrub Jays is impressive; it includes diving on the individual and sometimes pecking fiercely. Small-bodied snakes up to about 300 mm in total length are eaten.

Nest predation increases markedly throughout the nesting season. In March a Florida Scrub Jay nest has about a 70 percent chance of fledging young; in June the probability has plummeted to 30 percent. Because we have not noticed significant changes in the ways that helpers attend nests as the nesting season progresses (Stallcup and Woolfenden, 1978), we suspect the effect of helpers is simply swamped late in the breeding season as opposed to any important change in their helping performance. Therefore, we suspect that the suite of jay nest predators is more active in searching for jay nests, or is more successful at finding jay nests in late spring than in early spring. Snakes seem likely candidates, as their activities are known to be strongly seasonal.

These conclusions establish a major feature of Florida Scrub Jay demography, and it is a peculiar one from a theoretical point of view. Success breeds more success. Each year's reproductive effort by a given pair is *not* independent of past years' successes or failures. As we shall see in chapter nine, survival rates are similarly dependent upon one another, and upon past seasons' reproductive performance. This kind of interdependence between demographic variables that typically are treated independently raises complicated theoretical problems that we cannot deal with here. However, we emphasize the point because, to our knowledge, the phenomenon has never been addressed by demographers, much less modeled in detail in the literature on life history strategies.

Because the effects of helpers on the annual reproductive success of breeders are crucial and easily measured, they are prone to being overemphasized in the theoretical treatments of cooperative breeding. Virtually all such treatments, in identifying the costs and benefits to helping, focus on one or another version of a common value, the fractional increase in annual productivity that can be attributed to the presence of a helper. This value was called G by Brown (1975a, 1978a), $(N_2 - N_1)$ by Emlen (1978), and $(W_{AB} - W_B)$ by Emlen (1982b). In all these works (and others) this quantity is considered a central component to an individual's fitness. Corrected for the degree of relatedness between helper and offspring, this is the "kinship component" to an individual's inclusive fitness in these simple models. All these theoretical treatments thereby reduce the strategic comparisons of alternative strategies (values of which represent "fitness") to single-year comparisons, as follows: an individual should disperse and attempt breeding if this quantity is sufficiently low; it should remain and help if the quantity is greater than a certain level. Analogous, single-year criteria are used in assessing individuals' expected reproductive output with or without first undertaking a year of helping (e.g., Emlen, 1982a, b). We believe that these treatments consistently undervalue the fundamental observation that fitness, by definition, *must be measured as a lifetime parameter*. In comparing relative fitnesses of hypothetical, alternative strategies, it is not sufficient (and frequently will be misleading) to use annual values for reproductive parameters such as the ones derived in this chapter. The lifetime implications of each alternative strategy must be evaluated. We attempt to do this in chapter ten.

As a case in point, under certain circumstances we would expect cooperative breeding to exist even when "helpers" actually decrease the production of new young; this could occur for the following reasons. Helpers usually are the offspring of the breeders they attend. Mortality often is much higher during the first few months of a bird's life than later on. Therefore, because the older offspring have much higher reproductive value (simply *because* they have survived long enough to become helpers), they represent a greater contribution to parental fitness than do younger offspring, especially eggs and nestlings. The value of these older offspring as future reproductives can be enhanced further through "apprenticeship," territorial growth and budding, and a host of other features associated with gaining experience. Therefore, parents could be selected to tolerate older offspring in the territory despite a concomitant decrease in fledging success, given the appropriate demographic conditions. Emlen (1982b, p. 43) recognized this special case, but in his model all benefits to the breeder must accrue during the offspring's first year as a reproductive (year 2 of life) rather than over the lifetime of that offspring. The concept of reproductive value is missing from Emlen's treatment, and this feature we discuss in the next chapter.

CHAPTER NINE

Survivorship and the Life Table

SYNOPSIS

Mortality of young fledglings is high. Survivorship increases continuously over the first 12 months of life, after which it approximates that of adults. Fledgling survival to age 1 year is 34 percent. First-year survival is not correlated with fledging weight, nor with any variable of family constituency or population density, but is positively correlated with breeder survival between years. Breeder survival averages .82 annually and is equal between sexes. Senescence may occur after age 16 years. The presence of helpers significantly lowers breeder mortality. Breeder mortality shows a slight peak during summer. Helper death rates are not equal between the sexes. Yearling females show highest annual mortality (.35), followed by older female helpers (.26), yearling males (about .20), and older male helpers (.16). Male helpers rarely practice dispersal forays, but females do so regularly. Consequently, mortality among male helpers approximates that of breeders with helpers, while female helper mortality is much higher. The Florida Scrub Jay survivorship curve shows most characteristics of Deevey's Type III curve including age-independent adult mortality, observed in our population through age 10 years. Modal age of breeders is 3 years, median age is between 5 and 6 years, and 20 percent of breeders are older than 10 years. The mean life expectancy for a fledgling is about 2 years. Stationary age structure of the population includes 31 percent nonbreeders at onset of the average breeding season. Yearlings form the commonest age class, with roughly exponential reduction in relative abundance there-

after. The $l_x m_x$ curve is more skewed toward older age classes than in any other studied passerine bird. Over half the offspring are produced by females older than age 5 years. Growth rate of the population is zero ($R_o = 1.0$, r = 0). Mean generation time is about 6.6 years. Reproductive value peaks at age 4 years, but remains nearly as high through age 14 years. Fecundity and survivorship are sufficiently high in later years to allow selection for strategies early in life that maximize opportunities for, and success at, breeding later in life.

Two age-specific demographic parameters—fecundity and survivorship—constitute central components of the life history strategies among iteroparous breeders (reviews by Ricklefs, 1973; Stearns, 1976; Horn, 1978). In the preceding chapter we established values for effective "birth rates" in our Florida Scrub Jay population and showed the extent to which this measure varies with breeder age, breeding experience, and particularly the presence or absence of helpers. To complete our analysis of the technical demographic features of the study population, we present in this chapter our data on survivorship and age structure, followed by a summary of the Florida Scrub Jay life table.

Especially when dealing with species exhibiting delayed reproduction, accurate estimates of survivorship during all phases of life are crucial to analyzing the selective advantages of alternative life history strategies. Life expectancy during the prereproductive period bears directly on the potential costs to fitness in delaying breeding for one, two, or even three or more years, as do many jays in our population (chapter four). Furthermore, as Brown (1969, 1978a) has emphasized, selection for delayed breeding almost always is a prerequisite for the evolution of helping behavior. Therefore, as measured by differing mortality rates, the cost of choosing between a "dispersal" versus a "stay-home-

252

and-wait" strategy ultimately determines whether a population will evolve toward cooperative breeding. The costs of delaying reproduction can be measured only relative to the expected benefits once breeding status is attained. These benefits depend in large part upon the survivorship curves of breeders.

With the above considerations in mind, we analyze data on Florida Scrub Jay survivorship in three parts, corresponding to the three main life phases: first year (juveniles), older prereproductive (helpers), and breeders. Because measuring survivorship among juveniles and breeders is relatively simple, we treat these categories first, before dealing with the problems of distinguishing between long-distance dispersals and deaths among older helpers.

9.1. FIRST-YEAR SURVIVAL

Juvenile Survival Rate

Because nearly all young jays remain in their natal territory at least one year after fledging, we have obtained a picture of juvenile survival that lacks the complications of postfledging dispersal so troublesome to most studies of avian demography (Nice, 1937, 1943, 1957; Nolan, 1978; Ricklefs, 1973). Mean fledging date in our population is about 15 May, and every surviving individual is censused thereafter, at mid-month, as long as it remains in the study tract. These monthly survival data for nine consecutive fledgling year classes are summarized in table 9.1. The resulting average survival curve is shown in figure 9.1. Over nine years, an average of 34 percent of 456 fledglings survived for at least one year.

Figure 9.1 shows the mean survival rates, converted to their annual equivalents, during periods of one, two, three,

TABLE 9.1. Survival of Juvenile Florida Scrub Jays to Age 12 Months (1970–1978)

Month	(No. Mos.)	Year Class									Total	Survivorship			Annual Survival Rate
		'70	'71	'72	'73	'74	'75	'76	'77	'78		Pooled	Mean	± S.D.	
May	(0)	41	40	25	42	51	70	56	58	73	456	1.000	—	—	—
June	(1)	29	32	18	28	29	45	32	43	50	306	.671	.678	± .076	.080[a]
July	(2)	23	26	18	27	28	42	26	39	46	275	.603	.610	± .076	.278
August	(3)	22	21	16	20	22	37	19	39	41	237	.520	.523	± .101	.168
September	(4)	22	19	16	19	18	36	17	35	38	220	.482	.489	± .109	.409
October	(5)	21	16	12	18	16	35	13	32	37	200	.439	.436	± .105	.319
November	(6)	21	14	8	18	16	35	13	30	35	190	.417	.406	± .105	.540
December	(7)	21	14	7	18	15	34	13	30	35	187	.410	.398	± .110	.826
January	(8)	20	14	7	18	14	32	13	29	35	182	.399	.382	± .99	.722
February	(9)	19	14	7	18	13	31	12	29	32	175	.384	.375	± .103	.625
March	(10)	18	14	7	17	13	31	12	27	32	171	.375	.366	± .094	.758
April	(11)	18	14	7	15	13	31	12	24	29	163	.357	.350	± .083	.563
May	(12)	18	13	6	14	13	31	12	24	28	159	.349	.339	± .087	.742
Proportion surviving:		.44	.33	.24	.33	.25	.44	.21	.41	.38	.349	—	.339[b]	± .087	—

[a] Annual survival rate = (monthly survival rate)[12].
[b] Mean annual survival (.339) calculated as arithmetic average of 9 annual mean survival rates.

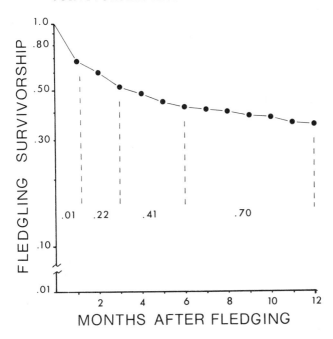

FIGURE 9.1. Mean survivorship of fledglings by month through their first year, from mid-May to mid-May. Dashes separate four periods lasting one, two, three, and six months, between which survivorship steadily increases. Average survivorship values for each period are shown, standardized to an annual rate.

and six months' duration, covering the first 12 months of fledgling life. As is typical of young birds (Ricklefs, 1973), survival rate increases dramatically with age during the juvenile period. This increase is illustrated by month in figure 9.2, where survival is again converted to annual rates. The increase is significant and steady over the first eight months. After 12 months young Florida Scrub Jays exhibit survival that is about 10 to 15 percent below the annual rate for breeders.

Two apparent dips in monthly survival of juveniles are shown as dashed lines in figure 9.2. These occur in August

255

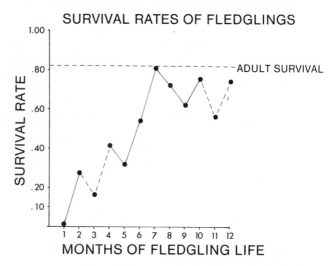

FIGURE 9.2. Monthly survival rates of fledglings to age 1 year, converted to a 12-month rate to compare to the annual survivorship of breeding adults (dashed line). Two substantial dips in the otherwise increasing survival rate (also dashed) correspond to periods when fledglings wander outside their natal territories.

and April, coinciding with the periods of wandering by both juveniles and older helpers discussed above (section 7.6. "Fledgling Wandering"). Although a few first-year birds may permanently leave the study area during these periods, we have evidence for this in only one case. We suspect that most of the added losses during August and April represent real increases in mortality associated with wandering through unknown neighboring territories during these two relict dispersal phases. Of the 159 known yearlings (table 9.1), only 14 (8.8%) were not in their natal territories 12 months after fledging. Three of these (all females) had paired nearby at age 12 months, and 9 more were either homeless or established as helpers in adjacent territories.

Individual Variation in Juvenile Survival

In contrast to survival of eggs and nestlings, the survival of juveniles once out of the nest is apparently independent of the presence or absence of helpers in the family. As shown in chapter eight, survival by fledglings of simple pairs was 36 percent as compared with 34 percent for fledglings in families with one or more helpers.

Also in the preceding chapter (see table 8.12), we showed that survival to yearling age is independent of the number of brood mates with whom a young jay leaves the nest. Larger fledgling broods produce more yearlings by exactly the predicted amounts if each fledgling is given a .34 probability of surviving one year regardless of the number of sibs also present in the territory.

Woolfenden (1978) showed that survival of fledglings from the 1970 through 1975 year classes through independence (age 3 months in that publication) was not correlated with nestling weight. This conclusion warrants reexamination and elaboration here. If correct, it suggests that beyond some critical minimum, the amount of food provided to nestlings by parents and helpers does not affect chances of surviving once out of the nest. Such a result would contradict well-established gospel about the importance of maximizing food intake by nestlings as a means of increasing their well-being as fledglings (e.g., Lack, 1968).

To test the effects of weight upon fledgling survival, we use survival data for 292 fledglings whose weights as nestlings at age 11 days are known. At this age nestlings have attained about 80 percent of their fledging weight (Woolfenden, 1978), so we assume that day 11 weight provides a good index of weight upon leaving the nest. Figure 9.3 graphs various measures of postfledging survival against

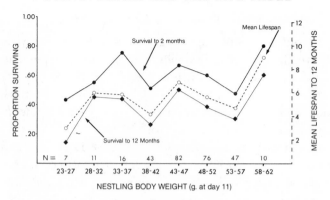

FIGURE 9.3. Fledgling survivorship in relation to body weight at banding date as a nestling. Shown for each four-gram increment are the proportions of the sample (N) surviving to age 2 and 12 months, and the mean life-span (where the maximum attainable is 12 months).

day 11 weights. Although a positive correlation is suggested by the extremes, where sample sizes are small, we can find no indication of a significant trend toward increased survival among heavier fledglings. A Spearman rank correlation of .03 was not significant, and additional tests of various permutations yield even less of a trend. We conclude that survival to independence (age 2 months here), or to age 12 months, is independent of body weight at day 11, and hence probably at fledging time. Again, it appears that the provision of extra *food* by helpers is not itself measurably beneficial to the nestlings that receive it.

In sum, we can find no significant correlate indicating that any category of fledglings has a better chance than any other of surviving to age 1 year. In the average year, each and every fledgling has about one chance in three of becoming a yearling. The story becomes more complicated, however, when year-to-year variations in survival are considered.

258

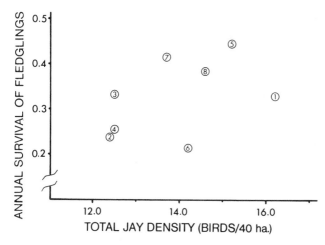

FIGURE 9.4. Annual survival of fledglings in relation to the density of jays in the study tract, averaged throughout the year. Circled numbers, 1–8, represent the years 1971–1978. No density estimate is available for 1970.

Annual Variations in Juvenile Survival

As shown above (table 9.1), survival over the first 12 months varies considerably from year to year, from a low of 21.4 percent (1976) to a high of 44.3 percent (1975). A significant relationship between juvenile survival rate and total population density might explain this variation and would imply that juvenile survival is strongly density-dependent. No such relationship exists (figure 9.4). We measured jay density through each year by averaging that year's densities during four sample months—June, September, January, and March. The absence of a relationship persists even if we weight June densities more heavily than other months (because June is early in the postfledging period), or when density of only breeders and older helpers is used rather than total jay density. We can find no evidence for density-dependent juvenile mortality.

A startling relationship exists between the survival of

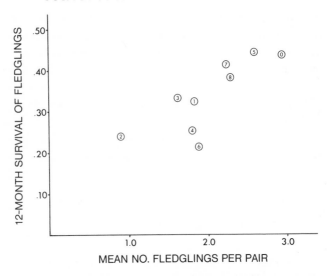

FIGURE 9.5. Correlation ($r^2 = .64$) between fledgling production and the ensuing survival of those fledglings to age 1 year. Circled numbers, 0–8, represent the years 1970–1978.

juveniles over a given year and the absolute number of juveniles produced to begin with, as shown in figure 9.5. In years when production of fledglings is high, the subsequent survival of those fledglings is also high. We will return to the ecological circumstances leading to good versus bad years below, but whether the answer is variation in predation rates, or some other factor, the relationship graphed in figure 9.5 makes intuitive sense. In short, good years for eggs and nestlings continue to be good years for fledglings.

The pattern is less easily interpreted when juvenile survival is broken down into preindependence versus postindependence life stages. Fledgling survival to age 60 days bears *no* relation to that year's fledgling production (figure 9.6a), whereas the relationship is strong and highly significant when survival *from* age 60 days to age 1 year is compared to fledgling production (figure 9.6b). Therefore, whatever factor or factors result in a correlation between nesting success and older fledgling survival do not seem to affect survival during the first 60 days out of the nest.

We interpret this puzzling anomaly as follows: as discussed above (see figure 9.2) mortality of newly fledged young always is extremely high. Many fledglings disappear within a few days of leaving the nest. Because the young at this stage are flightless, weak, and awkward, they succumb to a variety of hazards (e.g., storms and accidents) that have less affect on nestlings and independent juveniles. Furthermore, whatever the predation rate may be on nestlings and older young in a given year, it *always* is high for newly fledged young that perch near the ground, beg incessantly, and cannot escape from a variety of ground predators, both diurnal and nocturnal. Therefore, it appears that fledglings in their first few weeks take part in a life and death game of chance, from which the odds of surviving to age 60 days always are about 50 percent (figure 9.6a). In this regard, it is perhaps significant that annual variation in survival of fledglings to independence is much lower than that of nesting success or survival from independence to age 1 year (CV = 11 percent versus 40, 27 respectively).

We suspect that annual variations in nestling and fledgling survival result primarily from fluctuations in predation rates from year to year. The case for nestling survival is argued in chapter eight (section 8.6). Very little direct evidence for variable predation rates on fledglings can be mustered, except that we see virtually no indication that starvation ever afflicts young jays out of the nest. One piece of indirect evidence supporting the predation hypothesis is that annual survival rates of fledglings are correlated with survival

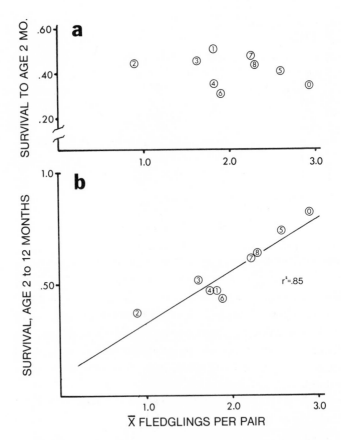

FIGURE 9.6. First-year survival of juveniles in relation to fledgling productivity. Survival to age 2 months (9.6a) shows no correlation, but survival from that point through age 12 months (9.6b) shows a strong one (r^2 = .85). Circled numbers, 0–8, represent the years 1970–1978. Linear regression line shown in 9.6b.

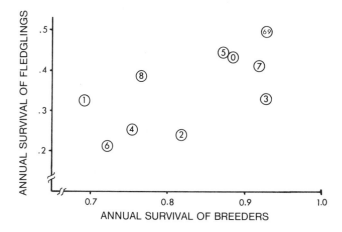

FIGURE 9.7. Correlation ($r^2 = .40$) between annual survival of breeders and their fledglings. Circled numbers, 69 and 0–8, represent the years 1969–1978.

rates of breeding adults during the same years (figure 9.7). Strong evidence exists that virtually all breeder deaths occur through predation. Although the relationship is not absolute, the correlation between juvenile and adult survival suggests that the same process is responsible for the bulk of the deaths in both groups.

9.2. SURVIVAL OF BREEDERS

Breeder Survivorship

Except in rare cases of divorce or departure following a mate's death (chapter four), breeders remain on the same territory from the time they first pair until their death. Breeder deaths become known to us as sudden, permanent absences from our monthly censuses, followed by the appearance in their territory of unpaired birds of the appropriate sex attempting to pair with the surviving widowed breeder. After assuming a breeder's death in our record books, we have never once been proven wrong by a later reappearance.

Table 9.2 summarizes the annual death rates of breeders by sex, using the same reproductive year (mid-May of one calendar year to mid-May of the next) as was used in calculating juvenile survival. For each year, the table shows the sample size of known breeders of each sex that could have survived for 12 months, the number of male and female deaths during those months, the annual mortality rates that result, and the proportion of the total sample of breeders that survived. Average annual mortality over 10 years was 18 percent, and the resulting .820 annual survivorship *is the same for both sexes.* This survival rate is produced either by averaging all 10 annual rates, or by pooling the entire sample of deaths (N = 94) and breeder years (N = 259).

The individual survivorship data (both sexes combined) are plotted in figure 9.8 on the customary semilogarithmic axes. We have 11 jays who were known as breeders in 1969 and hence could have been observed surviving 10 years (120 months) as breeders. The sample size increases steadily for shorter and shorter survival intervals, and a total of 142 breeders could have been observed surviving at least 1 year after their first recorded month as a breeder. Figure 9.8 shows that when calculated from a .820 annual survival rate, the proportions still alive after each survival period almost exactly fit the following negative function:

$$S_y = e^{-.0165y}$$

where S_y represents the proportion of some original sample surviving y months after first breeding. If y is measured in years the exponential constant becomes $-.1985$.

TABLE 9.2. Annual Survival Rates of Breeding Adults (1969–1978)

	Males			Females			
Year[a]	N[b]	Deaths	Annual mortality	N	Deaths	Annual mortality	Mean Survival
(1969)	(7)	(1)	(.14)	(7)	(0)	(.00)	(.929)
1970	17	1	.06	17	3	.18	.882
1971	26	7	.27	26	9	.35	.692
1972	32		.16	33	5	.15	.846
1973	27	2	.07	28	2	.07	.927
1974	28	9	.32	29	5	.17	.754
1975	27	2	.07	28	5	.18	.873
1976	30	11	.37	31	5	.16	.738
1977	31	2	.06	31	2	.06	.935
1978	34	8	.24	34	10	.29	.735
Total	259	48	.185	264	46	.174	$.820 \pm .091$[c]

[a] Each breeding year runs from 15 May of named year through 15 May of following year.
[b] Samples exclude five jays that died in mammal traps, resulting in unequal numbers of males and females in certain years.
[c] The .820 breeder survivorship excludes the 1969 survival rate, which is based on a small sample of breeders.

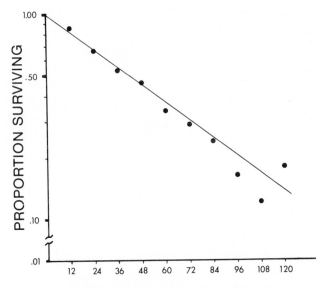

MONTHS OF LIFE AS A BREEDER

FIGURE 9.8. Observed survivorship of Florida Scrub Jays as breeders, including some unknown-age jays. Sample sizes vary from 11 individuals who became breeders early enough in the study to have lived 10 years (120 months) thereafter to 142 who could have been followed at least 1 year (12 months) after first breeding. Slope of regression line is .820, representing the average, age-independent annual survivorship of breeders.

Survival and Age of Breeder

A constant exponential decay in breeder survival (figure 9.8), the typical form for Deevey's Type III survival curve (Deevey, 1947), implies that death rate is independent of age. Starting with 1,000 breeders (of any age), a constant age-independent survival rate of .820 per year predicts that the last breeder will die 35 years later—an unusually long potential lifespan for a passerine bird. Our study still falls short of being able to test this prediction, but the constant decay rate shown in figure 9.8 suggests no drop in survival probability at least over the first 10 years of breeding.

If senescence occurs, it should cause an increased mortality rate among any sample of breeders containing the oldest age classes as compared with any sample that contains only the younger ones. During the period 1969–1973, our study tract was expanding and we were color-ringing a large number of unknown-age jays. Presumably these 67 original breeders represented an average cross section of all the age classes present in the population at large. All these breeders could have been observed to live at least 6 years (72 months) after first joining the study as marked birds. During the same period, from 1971 through the present, an annually growing percentage of the breeding population consisted of known-age birds who were banded in our tract as nestlings. The survival data for these two samples of breeders are shown in table 9.3 and graphed in figure 9.9. The early years' sample of unknown-age breeders appears to exhibit a lower survival rate than the known-age sample (about .79 versus .84 per annum), although comparisons of pair survivorship by year class are significantly different only at one point. Because the original sample of breeders includes a few representatives of the very oldest possible age classes, whereas the oldest of our known-age sample are only age 10 years, the mean ages of the two samples are quite different. This would make no difference in their survival rates if breeder survival were indeed age independent.

We interpret the lower survival rate among the original sample of breeders as preliminary evidence for an increased death rate, or senescence, at some age greater than 10 years. Using our estimates of the stationary age distribution among breeders (see below, especially table 9.8), we

TABLE 9.3. Survivorship of Known-Age and Original Unknown-Age Breeders

Years as Breeder[a]	Known-Age Sample			Unknown-Age Sample			χ^2
	N	No. Surviving	Proportional Survivorship	N	No. Surviving	Proportional Survivorship	
1	66	57	.864	67	56	.836	0.20
2	59	37	.627	67	40	.597	0.12
3	45	29	.644	67	29	.433	4.83*
4	37	20	.541	67	25	.373	2.72
5	29	12	.414	67	21	.313	0.90
6	24	9	.375	67	17	.254	1.27
7	17	6	.353	63	12	.190	2.03
8	8	3	.375	51	7	.137	2.78
9	—	—	—	36	4	.111	—
10	—	—	—	14	2	.143	—
$\overline{X}$ Annual Survivorship			.84			.79	

NOTE: See text for distinction between these samples.
[a] Years as breeder are counted since *first recorded* year with breeding attempt; years measured from May to May.
*This difference significant ($p < .05$).

calculate that a precipitous drop in survival after breeders reach age 16 years would account for the difference in survivorship curves shown in figure 9.9. More likely, of course, breeders may simply show gradually increasing annual mortality beginning sometime before that age, with a few birds living well beyond it. As modeled by Botkin and Miller (1974), this would eliminate all breeders from the population by about age 20 years. In any case, although our sampling technique requires many more years' data for confirmation, it appears that senescence characterizes the tail end of the Florida Scrub Jay life table, a feature that has been demonstrated for only a few species of birds (see Botkin and Miller, 1974; Coulson and Wooler, 1976; Bulmer and Perrins, 1973).

Effects of Helpers on Breeder Survival

A frequently cited advantage to living in groups is that predation rate is lowered by the presence of additional pairs of eyes to detect and signal the approach of predators (Moynihan, 1962; Lazarus, 1972; Pulliam, 1973; Bertram, 1978). Among Florida Scrub Jays, which fall prey to a variety of diurnal and nocturnal predators, group living could confer a real advantage, both to breeders and to helpers, if longevity is indeed higher in larger families.

Table 9.2 showed that 94 breeder deaths occurred in 523 breeder years. Table 9.4 splits these deaths into samples of breeders with helpers versus those without. The respective annual death rates, 15.3 percent versus 23.2 percent, are significantly different ($\chi^2 = 4.89, p < .05$). Breeders living as a simple pair die about 1.5 times the rate of those living in groups of three or more. Larger samples are required to examine the effects of group size, or of potential biases owing to age differences, within this simple pair-wise test. The implications of this result, however, are profound.

269

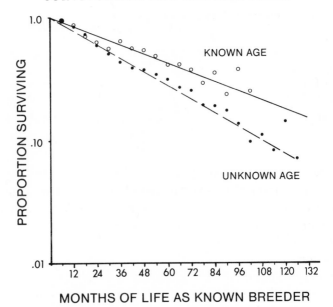

FIGURE 9.9. Survivorship of two samples of breeders. Original sample of unknown-age breeders (N = 67; closed circles) shows significantly lower survivorship than sample of known-age breeders (N varies from 8 to 64; open circles). Senescence may account for the disparity.

Increased longevity of group members over solitary individuals provides another way in which helpers can really "help" the breeders they live with. Furthermore, this help presumably is reciprocal, and provides a selfish, adaptive incentive for nonbreeders to remain with their natal group while seeking breeding status. This advantage must be taken into account in any overall analysis of the evolution of helping behavior (see chapter ten), and should be looked for among other group-breeding species (for an alternative view, see Koenig, 1981).

Seasonal Patterns in Breeder Mortality

Figure 9.10 illustrates the distribution of male and female breeder deaths through the year. In no two-month

TABLE 9.4. Effects of Helpers on Breeder Mortality

	Breeders with No Helpers	Breeders with ≥ One Helper[a]	Total
N (breeder years)	177	346	523
Deaths (D)	41	53	94
Mortality (D/N)	.232	.153	.180

[a] The with-helper sample includes only those breeders with helpers for 4 or more months during the 12-month breeder year.

interval is mortality statistically different between the sexes, further substantiating the observation of equal mortality throughout their lives *as breeders*. Heaviest mortality falls within the 60-day period between about 15 May and 15 July. Of the 94 breeder deaths, 27 (29%) occurred in this period as compared with a mean of 13.4 (14%) during all other two-month intervals ($\chi^2 = 11.32$ with 5 d.f.; $p < .05$). The period from May to July includes both the termination of the nesting season (see figure 8.1), when most pairs are feeding dependent fledglings, and also the onset of the definitive prebasic molt (Bancroft and Woolfenden, 1982), when flying ability is at its lowest annual level because of feather wear and loss. We suspect that these features together result in an increased susceptibility to predators. In contrast to the notion that breeding season mortality is relatively light (Krebs, 1970), it appears that late breeding season is the *most* dangerous time for Florida Scrub Jay breeders. Combining this observation with the potentially long life-span of a breeder (figure 9.8) suggests a possible explanation for the unusually brief and synchronous breeding season of the Scrub Jay in Florida. As the risk of mortality increases during the period of dependent young and feather loss, it may pay individual breeders to curtail any further nesting activities at this period rather than to attempt multiple broods. Additional nestings might protract the high-risk period to the point that expected longevity is reduced at the expense of future years' breeding

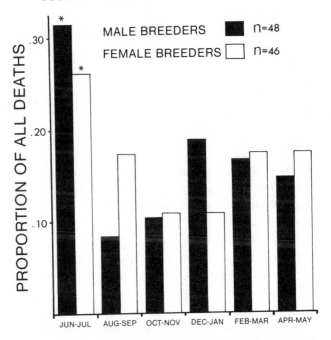

FIGURE 9.10. Seasonal patterns of mortality for male and female breeders. Both sexes show significantly higher mortality during June–July (asterisks) than during any other two-month period.

efforts. With a high annual probability of surviving from one breeding season to the next, more is gained by keeping each season's high-risk period as short as possible.

Effects of Older Helpers on Survival of Neighboring Breeders

On several occasions we have seen a helper make incursions into a neighboring territory and challenge the resident breeder of like sex a short time before that breeder disappeared. (As discussed in section 2.4, we can assume that a breeder who disappears from its territory is dead.) These interactions between helper and neighboring breeder, which have occurred in both sexes, suggest that sometimes a helper may become a breeder by somehow causing the death of the jay it replaces. Certainly we have seen sick or injured jays lose their territories and mates to invaders well before disappearing entirely (e.g., Woolfenden, 1976b; also section 4.2). It is even conceivable that some helpers become breeders by ousting healthy neighbors. Regardless of the causes and processes, if helpers do occasionally evict breeders the disappearance of breeders should occur more frequently during years with higher densities of older helpers. We tested this idea by comparing the death rates of breeders from one nesting season to the next (mid-May to mid-May) with density of older helpers in June of the same year (figure 9.11). The resulting strong correlation suggests that at least one density-dependent interaction may affect adult survivorship. At this time we can only guess at its mechanism, as many independent factors may be involved. We are intrigued by the possibility, however, that older helpers may be actively competing for breeding vacancies within their neighborhood more than we had suspected previously.

9.3. SURVIVAL OF OLDER HELPERS

Survival of juveniles and breeders can be measured directly from our monthly census data, but the survival of helpers older than age 12 months cannot. We can safely assume that disappearances of juveniles and breeders represent deaths, but each disappearance of an older helper can represent either a death or a permanent, successful dispersal to a breeding territory outside of our study tract. While we have succeeded in finding a number of long-distance dispersers (see figure 7.3), we cannot possibly have located all of them. This lack of certainty about the fate of

273

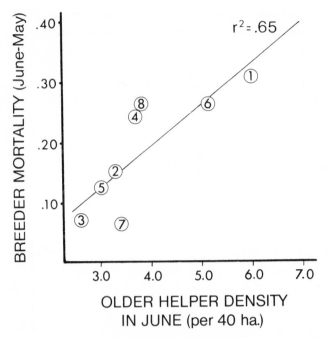

FIGURE 9.11. Correlation between breeder mortality and older helper density, suggesting that helpers may contribute to the demise of neighboring breeders. Circled numbers, 1–8, represent the years 1971–1978. No density estimates are available for 1970.

helpers is unfortunate. An accurate estimate of life expectancy among older helpers is essential in evaluating the potential costs and benefits of delayed breeding.

We approach the question of mortality among older helpers by asking what their *maximum* death rates could be, *then correcting these rates downward* by some amount approximating the rate at which older helpers successfully dispersed out of our tract. Maximum death rates simply equal the proportion of helpers (of a given sex and age class) that disappeared from the study tract between successive May censuses without becoming breeders. Table 9.5 shows the

274

LE 9.5. Fates and Death Rates of Yearling and Two-Year-Old Helpers

Age Class	N	Breeders[a]	Fates One Year Later			Death Rate	
			Breed	Help	Disappear	Maximum[b]	Actual
lings (1970–1978)							
es	64	252	23	32	9	.141–.225	.22
ales	81	252	21	26	34	.420–.466	.35
nown sex	7	—	0	0	7	—	—
year olds (1971–1978)							
es	28	235	9	12	7	.250	.15
ales	24	235	14	1	9	.375	(.34).26[c]

mber of breeders allows calculation of how helper density compared with breeder sity; see text and Appendix M for the use of this ratio in calculations of helper mortality. ximum death rates calculated by dividing number of disappearances by total number elpers for each age and sex class; actual death rates are lower because some disap- ances represent successful emigrations from tract; see Appendix M for calculations. 9.3. "Survival of Older Helpers" for explanation of this adjusted mortality estimate.

pertinent data on 152 yearling jays, and 52 two year olds. These data are broken down in more detail in Appendix M.

Death Rate of Yearlings: Calculation

Of the 152 yearling jays in our 1970–1978 sample (table 9.5), 7 were of unknown sex when they disappeared. Because we have a harder time positively establishing the sex of *male* helpers (section 2.4. "Field Procedures," regarding the female-specific "hiccup" vocalization), we suspect that most or all of the unknown-sex yearlings who disappeared were males. Because of the uncertainty, however, we must express our yearling death rate estimates as ranges, not as single figures. Among males, 9–16 yearlings disappeared out of 64–71 individuals, yielding maximum mortality of 14.1 to 22.5 percent annually. Among females the equivalent range is 42.0 to 46.6 percent.

To correct these estimates, we must arrive at assumed rates of successful dispersal from the study tract by male and female yearlings. We assume that our emigrant helpers entered the breeding population at large with the same age distributions already demonstrated *within* the tract (chapter four, table 4.1). Given these proportions, combined with a known breeder death rate of 18 percent and known ratios of yearling density to breeder density, we calculate the number of jays *expected* to have become breeders between ages 1 and 2 years. Subtracting from this the number *observed* to have done so in our tract yields an estimated number of helpers that must have become breeders outside the tract. Subtracting this number from the total number of disappearances gives the estimated number of deaths. Annual mortality is then calculated by dividing these deaths by the total number of yearlings in the sample. Figure 9.12 diagrams the calculations just described, using female yearlings as a sample group. The calculation assumes that immigration and emigration are equal in our study tract. Our best current estimates of mortality among yearlings (see Appendix M for calculations) are shown in table 9.5: about 35 percent of the females and as many as 22 percent of the males die during their second year of life.

Death Rate of Two Year Olds: Calculation

Mortality estimates among two year olds are not confounded by helpers of undetermined sex. The method of calculation is exactly the same as for yearlings (figure 9.12). Appendix M outlines the calculations, and the results based on 52 two year olds are shown in table 9.5. Our current estimates of mortality among these older helpers are about 34 percent for females and 16 percent for males.

It must be noted that our sample sizes—even after a decade—are marginal for calculating death rates in the two-year-old age class. Just a few individuals added to or subtracted from any column in table 9.5 would drastically

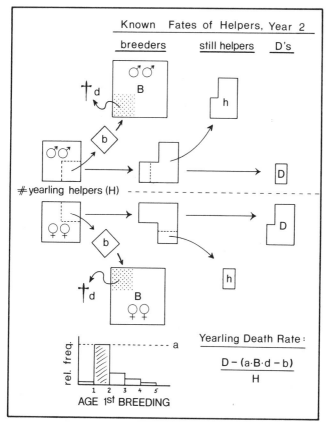

FIGURE 9.12. A method for calculating death rates of yearling help-ers in a stable population, given that some of the total disappear-ances (D) represent successful dispersal outside the study tract. A known proportion (d) of the breeding population (B) dies annually, leaving B•d breeding vacancies. If a is the proportion of first-time breeders that dispersed as yearlings, the total number of yearlings filling breeding vacancies is a•B•d. An observed number (b) of year-lings became breeders within the study tract, leaving an average of (a•B•d − b) that became breeders outside the tract. Subtracting these successful long-distance dispersers from the total number of year-ling disappearances and dividing by the total original sample of yearling helpers (H) yields the estimated death rate of yearlings. Note that for Florida Scrub Jays, females show a much higher yearling death rate than males.

alter the results. (For the same reason, death rates among helper age classes older than two years are impossible to calculate as yet.) Appendix M shows that of the nine female two year olds that disappeared, five did so in one year (1977), which was two years after an unusually successful breeding season produced a peculiarly large number of females. Removing this age class from the mortality analyses has little impact upon the yearling death rate estimates, but reduces the estimates among two year olds to *about 18 percent*. The estimate of 26 percent annual mortality among two-year-old females shown in table 9.5 corrects this bias by splitting the difference between 18 percent (excludes 1977) and 34 percent (includes 1977).

Patterns of Helper Mortality

Table 9.5 summarizes an extremely important trend in helper mortality. Females, especially the yearling females, suffer drastically higher annual mortality than do males. Indeed, mortality of yearling females is considerably higher than that of juveniles (including juvenile females) near the end of their *first* year of life (see figure 9.2). Although mortality among female helpers older than yearlings appears to drop somewhat, it probably remains significantly higher than that of males *at any age*.

The great disparity between mortality rates of male and female helpers clearly arises from their differing dispersal strategies. As described in chapter seven, yearling females frequently leave home for extended periods, and during these forays they appear to wander from territory to territory attempting to join other jay families. This behavior results in earlier and longer-distance dispersals than those of males. It also exposes the females to a variety of environmental dangers that are minimized by remaining on familiar ground among a group of familiar compatriots.

278

The cost of this active, extraterritorial searching strategy is an annual risk of death about twice that experienced by all the other, more sedentary categories of adult Florida Scrub Jays.

Male helpers, who seem to leave home only rarely (figure 7.4), exhibit a death rate roughly comparable to that of breeding adults *with helpers* at approximately 15 percent per year (cf. tables 9.4, 9.5). This substantiates our speculation (section 9.2) that reduced mortality among breeders who live in groups of three or more is reciprocally experienced by any of the helpers remaining in these groups. Reduced mortality certainly represents a critically important advantage to the nonbreeding individual in remaining at home. It is a demonstrably safer place to live.

The death rates just noted suggest that breeders suffer no appreciably greater mortality than helpers who remain at home, even though helpers do not build nests, lay eggs, or tend clutches. Therefore the "cost of breeding," compared to simply living on territory, appears to be negligible if measured strictly in terms of survival potential.

The mortality rates just calculated are the first such estimates for breeding-age helpers in any cooperative bird. The sample sizes are just barely sufficient to make general statements about helper survival, even after 10 years' accumulation. Because differences in survival probabilities of just a few percentage points can have profound influence on life table characteristics, this critical period in the life history of the jays represents our weakest link in assembling the complete survivorship curve. Additional years' data will strengthen our confidence in the precise values for age- and sex-specific survivorship of helpers, and also will allow us to attach standard error estimates to the values obtained. By now, however, we are confident that the overall differences between male and female survivorship, and the sim-

279

ilar mortality values for male helpers and breeders, will not change appreciably from our present estimates.

9.4. LIFE TABLE AND AGE STRUCTURE

The Survivorship Curve

Each year of our study, we begin to monitor the survival of a new cohort of young jays within the study tract. As discussed above, nearly all surviving members of these cohorts remain in the tract for the duration of their lives, first as helpers and later as breeders. Therefore, we can calculate the proportion of each cohort that survived to year x, as long as x is less than, or equal to, the number of years between that cohort's fledging and the present. This provides an observed survival curve for our population. We have followed 1 cohort since 1969 (10 years) and 10 different cohorts for at least 1 year postfledging. Thus our survivorship curve reaches only to 10 years, and our most reliable estimates lie during the first few years, where sample sizes are ample. Figure 9.13 shows our observed survivorship curve, with open diamonds showing the portion of the curve represented by original samples of fewer than 100 individuals. Points in figure 9.13 represent survival proportions at 2-month intervals. The data are summarized into 1-year intervals in table 9.6.

From ages 3 through 10 years, the relatively constant slope of the curve in figure 9.13 corresponds to the annual .820 survivorship already derived for breeders (table 9.2). We have virtually no way of estimating how the curve looks beyond age 10 years (but see section 9.2 regarding possible senescence). It should be noted that we see no evidence of steadily increasing mortality during this period (contra Botkin and Miller, 1974).

TABLE 9.6. Observed 10-Year Survivorship of Yearling Jays in Study Tract

Years Alive Postfledging	N	Sample Surviving[b]	Survivorship	
			Postyearling	Postfledging
1	156[a]	156	1.000	0.339[c]
2	156	102	.654	.222
3	132	71	.538	.182
4	119	51	.429	.145
5	85	36	.424	.144
6	68	24	.353	.120
7	55	16	.291	.099
8	47	10	.213	.072
9	35	7	.200	.068
10	11	3	(.273)[d]	(.093)[d]

[a] Sample of yearlings in this table (N = 156) differs from that in table 9.1 (N = 159) as follows: this sample excludes 1978 cohort (N = 28), and includes 1969 cohort (N = 8) plus jays banded early in the study as known-age juveniles but not as nestlings (N = 17).
[b] Samples of known survivors exclude dispersed jays known to be alive outside the study tract; therefore survivorship figures here are minimum estimates.
[c] Postfledging survivorship to age 1 year (.339) taken as annual mean shown in table 9.1.
[d] Parentheses enclose survivorship values based on small sample sizes.

Figure 9.13 is actually an underestimate of absolute survival rates, primarily because some small fraction of the nonbreeding group *did* survive as breeders outside our tract, thereby slightly reducing the observed survival of marked jays. Furthermore, it is desirable for comparative purposes in demography to plot survival estimates beginning with the earliest stage of life—in the present case, the egg (Deevey, 1947; Hickey, 1952). For these reasons, we calculated a synthetic curve for the Florida Scrub Jay, using in composite fashion our previously derived values for egg and nestling survivorship (chapter eight), juvenile survivorship (section 9.1), estimated helper survival (section 9.3), and breeder survival (section 9.2). These figures are listed in

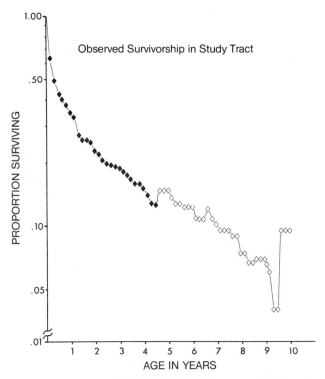

FIGURE 9.13. Observed survivorship of Florida Scrub Jays within the study tract, from fledging through age 10 years. Closed diamonds represent sample sizes of more than 100 potentially surviving individuals. Curve becomes less reliable at the end because of small sample sizes. Males and females combined.

table 9.7, where the synthetic survivorship schedule is projected through age 31 years. The resulting survivorship curve is graphed in figure 9.14. Because males and females suffer different mortality rates as helpers (table 9.5), we calculated their survivorship separately after age 1 year postfledging.

If senescence does occur, then the tail of the survivorship curve shown in figure 9.14 is artificially long. At least through

TABLE 9.7. Synthetic Survivorship Table of the Florida Scrub Jay

Life Stage	Age Interval Days	Age Interval Years[a]	Prop. Alive at Beginning (l_x)[b] Male	Prop. Alive at Beginning (l_x)[b] Female	Prop. Dying in Interval (d_x)[c] Male	Prop. Dying in Interval (d_x)[c] Female
Egg	0–6	0.000	1.000		0.210	
	6–12	0.016	0.790		0.055	
	12–18	0.033	0.735		0.082	
Nestling	18–24	0.049	0.653		0.080	
	24–30	0.066	0.573		0.090	
	30–35	0.082	0.483		0.072	
Fledgling	35–66	0.096	0.411		0.132	
	66–96	0.181	0.279		0.028	
Juvenile	96–127	0.263	0.251		0.036	
	127–158	0.348	0.215		0.014	
	158–188	0.433	0.201		0.022	
	188–219	0.515	0.179		0.012	
	219–249	0.600	0.167		0.003	
	249–280	0.682	0.164		0.007	
	280–311	0.767	0.157		0.003	
	311–339	0.852	0.154		0.004	
	339–370	0.929	0.150		0.006	
	370–400	1.01	0.144		0.005	
Yearling	400–765	1.10	0.139		0.031	0.049
Helper to	765–1130	2.10	0.108	0.090	0.018	0.019
Breeder	1130–1495	3.10	0.090	0.071	0.015	0.013
	1495–1860	4.10	0.075	0.058	0.014	0.011
	1860–2225	5.10	0.061	0.047	0.011	0.008
	2225–2590	6.10	0.050	0.039	0.009	0.007
	2590–2955	7.10	0.041	0.032	0.007	0.006
	2955–3320	8.10	0.034	0.026	0.006	0.005
	3320–3685	9.10	0.028	0.021	0.005	0.003
	3685–4050	10.10	0.023	0.018	0.004	0.003
	4050–5875	11.10	0.019	0.015	0.012	0.010
	5875–7700	16.10	0.007	0.005	0.004	0.003
	7700–9525	21.10	0.003	0.002	0.002	0.001
	9525–11350	26.10	0.001	0.001	0.001	0.001

[a] Age in years shown for beginning of age interval only; spaces in the columns occur where time intervals shift from 6 days to 30 days to 1 year to 5 years.

[b] Egg survivorship calculated from full sample of 1,337, which includes eggs preyed upon before complete clutch was laid (see table 8.11).

[c] Mean life-span of males: 389 days; females: 297 days, assuming no senescence. Sexes have similar l_x and d_x through age 400 days.

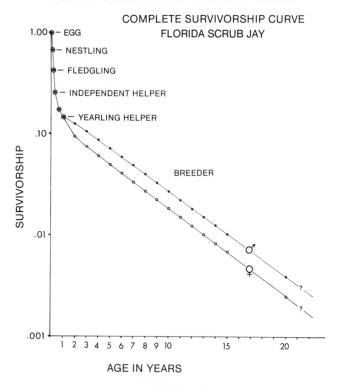

FIGURE 9.14. A complete, synthetic survivorship curve, from beginning of incubation through possible age of senescence (question marks). Males and females diverge slightly after age 1 year because of females' greater, dispersal-related mortality. Survivorship as breeders is identical between the sexes.

age 10 years, however, the curve shows the characteristic of Deevey's Type III survivorship (Deevey, 1947). Mortality is high during the early life stages, then levels off to a constant rate through most of the adult period. In comparison to other passerine birds for which such data are available, the Florida Scrub Jay shows significantly greater longevity (see Nolan, 1978; Ricklefs, 1973).

Age Structure: Breeders

Any population that maintains a roughly constant schedule of births and deaths eventually assumes a stationary age distribution (Lotka, 1922). As discussed in chapter three (see figure 3.18), the breeding population in our study tract remained remarkably constant between 1969 and 1979, despite significant fluctuations both in fecundity and in mortality of the various age classes. Overall, birth rate and death rate have been about equal. Assuming that this decade is representative of the long-term situation in the region, especially in that major population fluctuations do not occur regularly, then our data already can provide reasonable approximations of the stationary age distribution of Florida Scrub Jays in south central Florida. Here, we compare the observed distribution of known-age breeders in our tract with the age distribution predicted by a Leslie matrix simulation, using only the age-specific fecundity and mortality measures already derived.

By 1975 about half of our breeding jays were of known age (figure 9.15), and an additional fraction represented immigrants whose probable ages were known (nearly all dispersing jays are age 2 or 3 years). Figure 9.16 shows the age columns for each of the five successive breeding seasons, beginning with 1975. Progressing from left to right in figure 9.16, one can observe each known-age cohort move down the age column, first gaining numbers as nonbreeders become breeders, then diminishing in size as the breeders within the respective cohorts progressively die off. Above the bold line cross-secting the columns in figure 9.16, the unknown-age breeders represent known immigrants. We place them in their respective age classes according to the known distributions of ages at first breeding (chapter

285

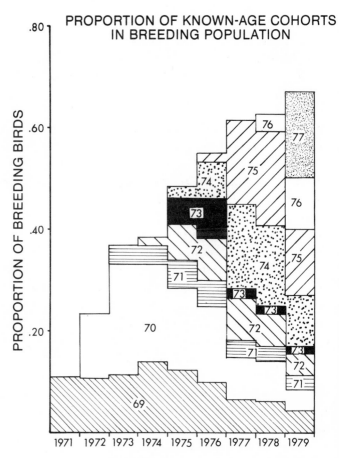

FIGURE 9.15. Proportions of known-age breeders within our study population, from 1971 through 1979 breeding seasons. Each cohort (numbers 69–77) first increases its representation, as helpers become breeders, then dwindles as breeders die. Total proportion of known-age breeders reached 68 percent in 1979 (and has stayed about the same for three breeding seasons thereafter).

four, table 4.1). We then average the proportions of the total breeding population represented by each age class over all years containing data on that age class. For classes above the bold line, we include the unknown-age immigrants in the sample. Only the known-age cohorts in age classes below the line were included, to provide a minimum estimate of their proportions during the earlier years. The data are summarized in table 9.8. The resulting average age column is shown at the right in figure 9.16 (also table 9.8, second from bottom row). These figures represent the *observed* proportions of breeders ages 1 through 10 years in an average Florida Scrub Jay population.

A semi-independent measure of the stationary age distribution among our breeders can be derived directly from the demographic parameters previously summarized, namely the breeder death rates and the age- and sex-specific recruitment rates (Lotka, 1922). Figure 9.17 graphs the *predicted* age distribution of breeders. The agreement between the two independently calculated age distributions suggests that our estimates of the actual stationary distribution in our population are reasonably close to the real distribution over the decade.

The following description of the breeding age structure of Florida Scrub Jays is based upon the data in figures 9.16 and 9.17:

1. The modal age of breeders is 3 years. This peculiarity reflects the level of delayed breeding characterizing our population, where virtually no yearlings breed and recruitment between ages 2 and 3 years exceeds the breeder death rate in the same interval.

2. Almost exactly half the breeders at any one time are age 5 years or less. Assuming that death rate after age 5 remains constant, the final jay out of a sample of 1,000 new breeders will die at about age 30 years. Using this projected

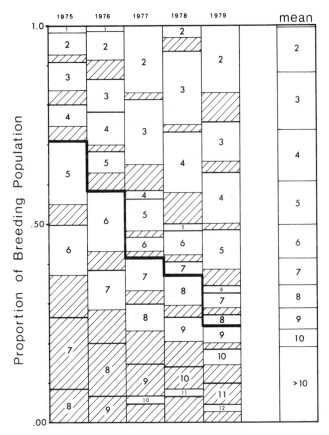

FIGURE 9.16. Age structure of the breeding population, 1975–1979, and the approximate average age structure through this period (right-hand column). Numbers, 1–12, identify separate cohorts within each column and show their age in years. Hatched areas within each cohort represent unknown-age jays, included within their minimum-possible age cohort. Above the bold line, these probably are accurate estimates of exact age. (See text and table 9.8).

Age of Breeders (Years since Fledging)

Year	1	2	3	4	5	6	7	8	9	10	11	12	N^a
1972	— .000	8 .125	7 .109										64
1973	— .000	2(2)[b] .074	12 .222	6 .111									54
1974	— .000	1(1) .036	2(3) .088	11 .193	8 .140								57
1975	1 .018	3(1) .072	4(2) .107	3(2) .090	9 .161	7 .125							56
1976	1 .017	4(3) .117	5 .083	5(1) .100	3(3) .100	9 .150	6 .100						60
1977	— .000	10(1) .184	10(4) .234	1 .017	5(1) .100	2(1) .050	5 .083	4 .067		1[c]			60
1978	— .000	2(2) .062	12(1) .204	10(5) .234	1 .016	4(1) .079	2 .031	5 .078	4 .063		1[c]		64
1979	— .000	12(5) .243	7(2) .129	9(1) .143	7(3) .143	1 .014	3(1) .057	2 .029	3 .043	3 .043		1[c]	70
$\bar{x}$ Prop. (Breeders)	.004	.114	.147	.127	.110	.084	.068	.058	.053	.043	?	?	485
$\bar{x}$ Prop. (Total population)	.003	.078	.101	.087	.076	.058	.047	.040	.036	.030	?	?	—

[a] N = Total sample of known-age and unknown-age breeders in study tract in given year. Proportions of this total are shown beneath each age class entry.

[b] Numbers in parentheses show known immigrants, listed under their minimum possible age, assuming they dispersed into the study tract at age 2 years.

[c] One known-age breeder, banded in 1967 before our study, is excluded from the averages in lower two rows to avoid biasing these mean proportions.

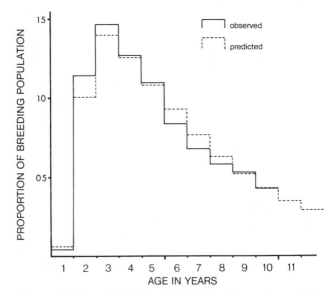

FIGURE 9.17. Age distribution of breeders predicted (dashed line) from breeder death rates and age- and sex-specific recruitment rates, compared with "observed" age distribution (solid line) calculated in table 9.8 and figure 9.16.

distribution, the mean age among Florida Scrub Jay breeders is about 7.1 years. This figure is considerably too high if senescence does affect older breeders.

3. Fully 20 percent of Florida Scrub Jay breeders are older than 10 years. Our oldest known-age breeder (Y–SA♀) died at age 13.6 years in December 1980. At present several jays are nearing her record. The Florida Scrub Jay is the longest-lived passerine species for which extensive data are available, although we suspect that other species, particularly among the Corvidae, will prove to have at least equivalent longevity.

Age Structure: Total Population

Table 9.9 shows the distribution of helpers, ages 1 to 7 years, in comparison with breeding-bird numbers between

1972 and 1979. Before 1972 many helpers were of unknown age, hence our decision to begin with that year in this analysis. Summing over all eight years (table 9.9, lower two rows) shows the average stationary proportions of the total population represented by helpers of each age class. Figure 9.18 graphs these proportions on top of the breeding age structure derived earlier. This "Christmas tree" graph, whose lower branches are heavily decorated with nonbreeding individuals waiting to become breeders, provides our best estimate of the overall stationary age distribution at the onset of the average breeding season in the Florida Scrub Jay.

9.5. $l_x m_x$ AND INTRINSIC RATE OF INCREASE

The average, relative contributions to total offspring recruitment by females of different ages are illustrated by the $l_x m_x$ curve, a distribution that sums to 1.0 in a completely stable population. The pertinent data for calculating the $l_x m_x$ schedule of our Florida Scrub Jay population are presented in life table format in table 9.10, where m_x (birth rate or fecundity) is measured in *fledglings produced per living female*. This measure may be more complex than in many bird species (Ricklefs, 1973; Hickey, 1952) because females first become breeders anywhere between the ages of one and five years, and because novice female breeders produce fewer offspring than do experienced ones (chapter eight). Table 9.10 shows the relevant proportions of surviving females breeding at each age (derived from table 4.1 and table 9.5) and incorporates the appropriate fecundity levels for first breeders and experienced breeders.

It is important to note that the m_x values shown in table 9.10 are not equivalent to the F_x fecundity values used by Leslie (1945, 1948, 1966) and others in calculating other demographic parameters. The simplest conversion of our

Table 9.9. Age Distribution of Helpers in Relation to Total Population

Year	Helper Age in Years							Total No. Helpers	Total No. Breeders	Total
	1	2	3	4	5	6	7			
1972	15(1)	8(1)	4(2)	2(2)	—	—	—	29	64	93
1973	8	8(1)	2	2(1)	1(1)	—	—	21	54	75
1974	9	5(1)	5	2	1(1)	1(1)	—	23	57	80
1975	18	2	1	—	—	—	—	21	56	77
1976	30	4	1	1	—	—	—	36	60	96
1977	13	9	1	—	—	—	—	23	60	83
1978	23	3	2	—	—	—	—	28	64	92
1979	29	6(1)	—	—	—	—	—	35	70	105
Total ♂ helpers	69	25	13	7	4	1	0	119	—	705
Prop. ♂ helpers	.098	.036	.019	.010	.006	.002	.000	.169	—	—
Total ♀ helpers	76	20	3	0	0	1	1	101	—	705
Prop. ♀ helpers	.108	.028	.004	.000	.000	.001	.001	.143	—	—
Total helpers	145	45	16	7	4	2	1	220	485	705
Prop. of population	.206	.064	.023	.010	.006	.003	.001	.312	.688	—

NOTE: Numbers in parentheses represent helpers whose ages were estimated.

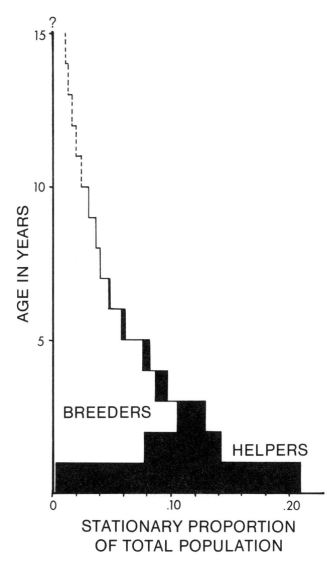

FIGURE 9.18. Stationary age distribution of a Florida Scrub Jay population at the onset of an average breeding season. Dashed line and question mark signify our incomplete knowledge of mortality rates among the oldest age classes.

TABLE 9.10. The Florida Scrub Jay Life Table (Female)

Age	l_x	d_x Breeder	Helper	Proportion Breeding Novice	Experienced	Fecundity Novice	Experienced	m_x	$l_x m_x$
0	1.000	—	0.66	0.000	—	0.000	—	0.000	0.00
1	0.339	0.18	0.36	0.040	0.000	0.786	—	0.031	0.01
2	0.210	"	0.26	0.500	0.040	"	1.000	0.433	0.09
3	0.172	"	0.26	0.366	0.540	"	"	0.828	0.14
4	0.140	"	0.26	0.084	0.906	"	"	0.972	0.13
5	0.115	"	—	0.010	0.990	"	"	0.998	0.11
6	0.094	"	—	0.000	1.000	—	"	1.000	0.09
7	0.077	"	—	"	"	—	"	"	0.07
8	0.063	"	—	"	"	—	"	"	0.06
9	0.052	"	—	"	"	—	"	"	0.05
10	0.043	"	—	"	"	—	"	"	0.04
11	0.035	"	—	"	"	—	"	"	0.03
12	0.029	"	—	"	"	—	"	"	0.02
13	0.023	"	—	"	"	—	"	"	0.02
14	0.019	"	—	"	"	—	"	"	0.01
15	0.016	"	—	"	"	—	"	"	0.01
16	0.013	"	—	"	"	—	"	"	0.01
17	0.011	"	—	"	"	—	"	"	0.01
18	0.009	"	—	"	"	—	"	"	0.00
19	0.007	"	—	"	"	—	"	"	0.00
20	0.006	"	—	"	"	—	"	"	0.00

. . .

life table to Leslie's format would involve indexing the age classes exactly as we have done (i.e., ages at the mean fledging date of 15 May). In this case, the first survivorship value, l_0, would represent the *yearling* age class, rather than the fledgling age class as in table 9.10. Setting $l_0 = 1$ requires dividing all subsequent l_x values by the first-year survivorship value, which is .339. All m_x values are then multiplied by .339 to represent the number of yearlings produced, rather than numbers of fledglings. The $l_x m_x$ column, of course, remains unchanged, although its first term is eliminated. This conversion has the disadvantage of discarding all information about juvenile survival. Alternatives, such

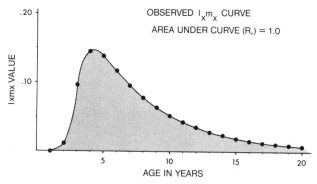

FIGURE 9.19. Florida Scrub Jay survival and fecundity schedules, combined into an $l_x m_x$ curve. Age classes 3–10 years contribute the bulk of the young jays recruited each year. In this absolutely stable population, the area under the curve (shaded) exactly equals 1.0 (see also table 9.10).

as picking up the initial age class some months after fledging but before age 1 year, require indexing all age classes from a time other than peak breeding season, an exercise we leave to the interested reader.

The $l_x m_x$ schedule presented in table 9.10 is graphed in figure 9.19. In comparison with all other passerines for which such data are available (see references in Ricklefs, 1973), the $l_x m_x$ curve of the Florida Scrub Jay is highly skewed toward older age classes. The curve illustrates the combined features of low and age-dependent fecundity, high adult survivorship, and delayed reproduction characterizing this race. All of these classically "K-selected" attributes (MacArthur and Wilson, 1967) are intimately involved in the evolution of cooperative behavior between offspring and parents, as discussed in the next chapter.

r *and Mean Generation Time*

In the core area of our study tract, rates of immigration and emigration have been virtually identical over 10 years

295

(unpublished calculations). This is reflected in the breeder density graph shown in figure 3.18. We used this evidence for population stability in our calculations of nonbreeder mortality in a preceding section of this chapter. As a result, it comes as no surprise that net reproductive rate, R_0, as calculated from our life table (table 9.10) is precisely 1.0 ($\sum_{x=0}^{20} l_x m_x = 1.0$; figure 9.19; see Michod and Anderson, 1980, for comments on a frequently committed error in calculations of R_0 and r based on field data). According to our models of population demography, the average Florida Scrub Jay exactly replaces itself. In such a population the intrinsic rate of increase, r, is zero. Based on our information on habitat saturation (chapter three), we strongly suspect that this zero growth is true for Scrub Jay populations throughout their range in Florida.

Unfortunately, the theoretical literature on population dynamics largely ignores the special case where $R_0 = 1.0$ and $r = 0$ (e.g., Ricklefs, 1979; Emlen, 1973). This void will plague us in chapter ten, where we attempt to compare fitnesses of individuals that use hypothetical, alternative strategies for dispersal and reproduction, thereby exhibiting different life tables in a model population *prohibited from growing*. We mention the point here because, given the life table, we are equipped to calculate our population's mean generation time, $\overline{T}$, a variable that classically has been linked to growth potential (Cole, 1954; Laughlin, 1965). In the classic logistic equations for population growth, mean generation time is indeterminate where r equals 0, and its importance in fitness comparisons is a matter of some dispute. Our estimate of this parameter will be used primarily in chapter ten, but we present it here for comparison to other bird species as one measure of overall population turnover.

We use the equation derived by Leslie (1966) for mean

generation time in an age-structured population of iteroparous breeders,

$$T = \int_0^\infty xe^{-rx} \, l_x m_x dx.$$

In our case, because $r = 0$, the result will exactly equal the
alternative expression, T_c, for cohort generation time or
the average age of mothers at the birth of their daughters,

$$T_c = (1/R_0) \int_0^\infty xl_x m_x dx$$

(see May, 1976, for elaboration on these and similar for-
mulas for estimating generation time).

Using the above equation and integrating over the 20-
year synthetic life-span shown in table 9.10, $\overline{T} = T_c = 6.60$
years. Integrating to infinity, the result is $\overline{T} = T_c = 7.26$
years. Given that senescence probably occurs (figure 9.9)
and that $R_0 = 1$ after 20 years, we assume that the former
estimate is the more accurate.

Reproductive Value

The effects of delayed breeding and high juvenile mortality show up directly in our calculation of reproductive
value, V_x, of individual Florida Scrub Jays. This measure
depicts the age-specific expectation of future production
for individual females, calculated according to Fisher's (1930)
equation, modified for the special case where R_0 is 1.0:

$$V_x = (1/l_x)\overset{\infty}{\Sigma} \, l_x m_x$$

The curve of age-specific reproductive values is graphed
in figure 9.20, again using the 20-year life-span convention
of our life table (table 9.10). The general form of the curve
does not change appreciably if either infinite potential longevity (slightly lower peak and longer tail to distribution)
or senescence (slightly higher peak and earlier truncation)
is assumed instead of the 20-year maximum life-span shown
here.

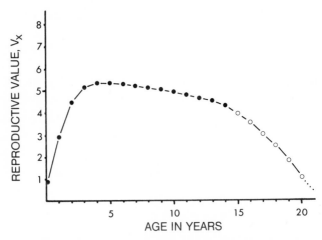

FIGURE 9.20. Approximate shape of the reproductive value curve for Florida Scrub Jays. Open circles beyond age 15 years signify our uncertainty regarding mortality rates at those ages. Senescence beyond age 15 would curtail the curve dramatically at that point and would lower the absolute values preceding it.

The most interesting features of the reproductive value schedule are its delayed peak (maximum value = 5.34, age 4 years) and its extremely gradual decline over the next 10 years. The peak value is attained only after all females become reproductive. The steeply ascending curve over ages 0 to 3 suggests that self-preservation and "self-improvement" strategies are favored early in life, extending over a longer period than in most other passerine birds, which breed earlier. Remaining in the natal territory and enjoying the reduced mortality associated with group living certainly represent one such strategy. Helping behavior might well represent another, especially if the experience gained as an apprentice breeder increases helpers' later ability to produce offspring once they join the breeding ranks. Without the appropriate controls, we cannot test this possibility in our population, but some suggestive data to this effect are shown in table 8.22. Additional evidence supporting the idea that helpers improve through practice is reported by Lawton and Guindon (1981) for the Brown Jay and Gabaldon and Balda (1980) for the Pinyon Jay.

Biological reproduction has been viewed as the transferral of reproductive value from parents to offspring (Ricklefs, 1979). In this sense it is interesting that the long, shallow decline in V_x following its delayed peak means that for a long period of time breeding individuals retain essentially the same value (as contributors to future generations) as their reproductive offspring. Indeed, only at age 14 years does reproductive value dip below that of the average two year old, the minimum age at which breeding typically commences. This result simply reflects the relatively low and (through this age at least) age-independent mortality of breeders. It does point out, however, that only *beyond* this age can reproductive value be considered measurably transferred to progeny. Perhaps it is significant, therefore, that our estimation of the age at which senescence may occur (section 9.2) falls exactly at this point. Of course, few breeders reach this age, and it will take us many more years to test empirically Hamilton's (1966) idea that senescence evolves in response to relative reproductive values of breeders and offspring.

9.6. CONCLUSIONS

A number of features presented in this chapter tie together the Florida Scrub Jay life history picture. Whereas reproduction (direct or indirect) is the ultimate "goal" of any animal, survival is the most crucial proximate road to that end. Many characteristics of the jays' survivorship schedule suggest directions that selection might have taken in molding the social system toward cooperative breeding. We conclude our presentation of the data by highlighting a few of these directions. We are fully aware that the discussion to follow resembles a paradigm for the classically K-selected strategy as first described by MacArthur and Wilson (1967), despite recent warnings against such unrigorous integration of life history traits into a simple, deterministic model (Stearns, 1977; Wilbur et al., 1974).

Survivorship of juvenile Florida Scrub Jays is extremely low during their first few months out of the nest. Even after jays reach a level of ability we call "independence" (age 2–3 months), their mortality rate is four times that of breeding adults (cf. tables 9.1, 9.2). By the end of their first year of life, juvenile helpers' survival approximates, but still is below, that of breeders. The first year of life is a period of intense learning, visible to us in a multitude of ways, not the least of which involves simple foraging ability (DeGange, 1976). Certainly, the equally important, elementary strategies of predator avoidance and dominance behavior are acquired and practiced during this long learning period.

In marked contrast, survivorship is extremely high once breeding status is attained. Furthermore, breeder survival remains high for a very long time thereafter, and reproductive success increases with age and experience (for at least as long as we have measured it). Therefore, reproductive value of an average individual exceeds that required for self-replacement by the time the individual is three years old, and it remains roughly at this level for over a decade thereafter. In a stable population, exceeding self-replacement can be considered a definition of the successful reproductive strategy (e.g., Horn, 1978). Clearly, for an average Florida Scrub Jay, then, the selective "battle" is won merely by becoming established as a breeder. Most individuals lose.

For any bird species, once the initial period of juvenile survivorship is past, individuals still face the problem of obtaining a mate *and* the problem of surviving while executing the search. If the environment is maximally crowded with territorial and long-lived breeders, to disperse into it via indefinitely random search is the most dangerous of several possible alternatives. This is demonstrated by the heavy mortality encountered by yearling female Florida Scrub Jays (table 9.5), many of which follow such a strategy. If other, less expensive, strategies exist by which nonbreeding individuals can become breeders, we should not be surprised to find them arising in such a population.

The points thus far compel us to paraphrase portions of an often-used definition of "crowding tolerance." The Florida Scrub Jay social environment is such that selection should favor strategies that simultaneously (1) maximize survivorship of young jays; (2) improve the ability of prebreeders in competition for a limited number of breeding vacancies; and (3) minimize the risk of mortality during dispersal. We emphasize that a jay can afford to meet all three goals at the expense of early breeding, because life expectancy is high once these hurdles are crossed. However, we add a fourth feature to the optimal strategy to insure that this last requirement is met: (4) maximize survivorship once breeding status is attained.

Of course, additional adaptive strategies can be cited in the K-selected environment of a cooperative-breeding jay (see Brown, 1974, but also Brown, 1978a). In particular, we leave out of the above list any mention of optimality in reproductive schedules. We might, for example, expect a relatively low and fixed clutch size that does not drain the breeders' resources during any given year of breeding, both

301

because breeder life expectancy otherwise is high and because annual variation in offspring survival rates is high (see Schaffer, 1974; see below). However, except for their rather brief and synchronous nesting season, Florida Scrub Jays show no obvious trend toward the greatly reduced annual reproductive effort that characterizes many other long-lived birds (Ashmole, 1971). Instead, in our view at least, their major life history responses to crowding represent an integration of the four strategic elements enumerated above, all of which relate to survival and dispersal more than to reproduction. In chapter ten we develop this idea in more specific detail.

Among the problems recently raised concerning the shortcomings of modern life history theory is that alternative models often predict the same result (Stearns, 1977). One fine example is the question of evolution of iteroparity and long adult life-span. On the one hand, this combination of characteristics has been viewed traditionally as a key feature of K-selected organisms (MacArthur and Wilson, 1967; Pianka, 1970, 1972). In this view, parental investment should be heaped upon few offspring at each reproductive period, with many such periods expected in the lifetime of an average adult. The response thus represents an adaptation to *stable* environments, which are constantly at or near carrying capacity (Pianka, 1970). On the other hand, delayed breeding, iteroparity, and low annual reproductive effort have been shown by stochastic models to be favored responses to *high variability* in juvenile mortality compared to variation in that of adults (e.g., Cohen, 1966; Schaffer, 1974).

In this example at least, the two seemingly alternative approaches turn out not to be mutually exclusive. We have made a case throughout this book (see especially chapter three) for the stability of our population, for its constantly

302

TABLE 9.11. Annual Variation in Offspring Production and Breeder Survival, 1970–1978

	Breeding Season										
	'70	'71	'72	'73	'74	'75	'76	'77	'78	Mean ± S.D.	CV
Fledglings/pair	2.93	1.82	0.89	1.62	1.82	2.59	1.87	2.23	2.28	2.01 ± 0.59	29
Juvenile survivorship	.439	.325	.240	.333	.255	.443	.214	.414	.384	0.339 ± 0.87	26
Yearly production	1.29	0.59	0.21	0.54	0.46	1.15	0.40	0.92	0.88	0.72 ± 0.36	51
Annual survival of breeders	.882	.692	.846	.927	.754	.873	.738	.935	.735	0.820 ± 0.091	11

NOTE: See Epilogue for discussion of the peculiar 1979 breeding season and subsequent rates of mortality.

saturated breeding structure, and for the importance of these features upon survival and dispersal strategies of Florida Scrub Jays. For comparison, however, we present table 9.11, which summarizes annual variation in production and survival of offspring and in survival of breeders. The contrast is conspicuous. Because of fluctuations in fledgling production and juvenile survival, annual production of young surviving their first year varies nearly five times more than does adult survival over equivalent intervals (CV = 51 vs. 11 percent, respectively). As pointed out by Hirshfield and Tinkle (1975), high variance in juvenile mortality might frequently characterize intensely competitive environments. Certainly, highly variable juvenile mortality seems to hold for the Florida Scrub Jay. Whereas theoreticians may argue that one model or another represents a more accurate expression for the evolution of a given set of life history traits, our evidence suggests that in at least one case (for which data are relatively complete) the models need not be constructed as alternatives. Features exist in nature that are common to both regimes, and the effects appear to be additive in this case. Whether this is a general phenomenon or one specific to a few extreme populations such as ours remains to be explored in works more general than the present one.

CHAPTER TEN

Evolution of Florida Scrub Jay Sociality

SYNOPSIS

In a stable population, fitness is maximized by leaving the largest number of successfully dispersed offspring over a lifetime of reproduction, regardless of the timing of that reproduction. Failure to account for lifetime reproductive success has caused oversimplifications and conceptual errors in previous models in the cooperative-breeding literature. Lifetime fitness can vary according to interactions between a large number of demographic variables involving survivorship, fecundity, relatedness, and dispersal probabilities. We express some important features of these interactions algebraically, stressing that relative availability of breeding vacancies and the methods by which they can be created are population measures of overriding importance when modeling the conditions under which helping behavior should arise. The Florida Scrub Jays' helper system can be viewed as a combination of strategies in which dispersal is delayed until a breeding vacancy can be located outside the natal territory or created from within it. Such a strategy is most favorable under conditions of high juvenile and prebreeder survivorship and in uniform, sharply bounded habitats that are saturated with long-lived breeders. Kin selection relaxes the conditions under which helping is favored over early dispersal, but it is of relatively minor importance at the absolute levels of direct and indirect reproduction typical of birds. In the Florida Scrub Jay, the helper system as we model it would evolve under present-

day demographic conditions even in the absence of the indirect, kinship components to fitness. Scrub Jays in western North America disperse early, rather than help, probably because they enjoy a relatively high probability of successful first-year dispersal (perhaps about .7 versus .4 or less in Florida). More complex social systems, including multiple-pair territories and larger group sizes, can be caused by relatively minor demographic changes from the present condition in Florida Scrub Jays.

Among other aims, modern social theory attempts to identify the various forces responsible for the evolution and maintenance of complex social interactions. Despite numerous advances, a recent compendium reflecting the state of the art in this field (Markl, 1980) identifies numerous dilemmas that still obscure our understanding of even the simplest routes to cooperative behavior in nature. In the words of Markl, the Dahlem Conference editor (1980, p. 8), "This report testifies as to how far away biologists are from really understanding what went on and goes on in the evolution of animal sociality." Among the common threads throughout the Dahlem report are two related problems germane to our present work (although by no means unique to the study of animal cooperation): (1) ideas about social evolution must be cast in terms of testable, alternative hypotheses that specifically relate behavioral strategies to their respective rewards in terms of inclusive fitness; (2) real-world testing between any hypotheses that postulate selective benefits of one behavior over another *requires* some comparative measurements of fitness for the two behavioral patterns being compared.

In this chapter we summarize our views on the evolution of Florida Scrub Jay sociality and its relationship to other cooperative-breeding bird species, especially jays. We begin by addressing the above two problems (in reverse order) as they relate to analyses of cooperative breeding. We summarize the Florida Scrub Jay demographic regime by presenting an example of the data set required to make detailed fitness comparisons for alternative life history strategies in cooperative breeders (table 10.1). We discuss some hypothetical alternatives open to nonbreeding jays, concentrating on the relationships between a few, crucial demographic parameters. By means of a simple algebraic model thereby generated, we address the question of kin selection in the evolution of avian cooperative-breeding systems. Finally, and most important, we bring all the above considerations together in a general, verbal hypothesis for the evolution of complex jay sociality. We close the chapter with some comments on what seem to us to be the most fruitful directions for future work in this field.

10.1. ESTIMATING FITNESS

In this and the following two sections, we are asking four basic questions. (1) Under what conditions should young birds of a territorial species elect to remain home instead of dispersing early to breed? (2) Given that they remain home, why should they help to raise young not their own? (3) How do the various demographic parameters affecting the above two questions compare in the magnitude of their influences? (4) Where do Florida Scrub Jays fit in with respect to our theoretical predictions about helping versus dispersing?

To answer the first three questions requires the formal construction of alternative strategies that share some common currency by which to compare their values. Strategy A will be favored over B if, and only if, A confers greater inclusive fitness upon its owner than does B. The fourth question can be addressed only by inserting actual or es-

307

timated values into the demographic variables within these equations. With sound theory and data, we should hope to show (1) why the particular population under study acts the way it does and (2) under what ecological circumstances we would expect it to change.

The most important life history measurements involved in cost-benefit analyses of cooperative breeding include: birth rates, age-specific survival probabilities, age-specific probabilities for successful dispersal, relatedness values between helpers and potential recipients, and some measure of how these values change with changes in group size. Ideally, actual values for each variable would be available, not just for the study population, but also for each alternative strategy and population we can conceive of. These values would permit direct comparison of different strategies entirely from field data, providing an ideal test of our hypotheses. Although no such ideal comparison is possible, we are luckier than most because Florida Scrub Jays provide statistical samples of several competing breeding styles in the same population. This variation allows direct measurement of certain values critical to the strategic analyses we will construct.

Table 10.1 presents a list of the parameters we consider crucial to the strategic analysis of any cooperative-breeding system, and it summarizes their respective values in the Florida Scrub Jay as calculated in chapters three through nine of this book.

The Currency

To estimate the inclusive fitness of individuals, and especially to compare two strategies, we are still limited to some predicted, average production of "fledgling equivalents" as our standard currency. In a heavily age-structured population a more ideal currency would be "breeder-equivalents" or "grandchild equivalents" instead of fledglings. These would more closely correlate with the actual replication of genome-equivalents throughout succeeding generations (the only true Darwinian fitness currency), but they remain unavailable in our sample of field data. We *can* produce accurate estimates of fledgling production by various strategies, and for the present we assume this to be a sufficient index of comparative fitness.

Given any pair of life history strategies, what end product should be examined when deducing which strategy will be favored? The question is a general one, faced by theoreticians for a generation (e.g., Cole, 1954; Pianka and Parker, 1975; Southwood, 1981; reviews by Stearns, 1976, 1977, and Horn, 1978). Curiously, it has been virtually ignored in theoretical treatments of avian cooperative breeding even though the term *inclusive fitness* is used repeatedly. Always implied is the existence of some measurable, "inclusive" end product that somehow "includes" everything, but what exactly *is* this end product? The question, and many of its complexities, was addressed by Brown (1980, pp. 122–125), but only in a general way. We believe that failure to pay strict attention to this theoretical issue has led to conceptual errors and misinterpretations of field data in the literature on cooperative breeding.

Under conditions of population growth or decline, and in cyclic or highly unstable environments (MacArthur, 1968; Schaffer and Tamarin, 1973; May, 1977), fitness depends not only on reproductive rate but also on time between succeeding generations (Pianka and Parker, 1975; Southwood, 1981). These are the conditions most often assumed (frequently implicitly) in life history theory. In these examples, fitness of a strategy can be equated to the magnitude of r, the rate of growth for a population following that strategy. Because r is inversely proportional to mean

TABLE 10.1. Some Important Demographic Variables among Cooperative Breeders and Their Average Values for Florida Scrub Jays

Description	Our Name for Variable	Observed Value in Florida Scrub Jay	Reference
Survival (L, l)			
Fledgling to earliest possible dispersal age	L_o	0.49 (age 4 mo.)	Table 9.1
Fledgling to age 1 year within natal territory	L_j	0.34	Table 9.1
As juvenile helper ($=Lj/L_o$)	L_1	0.69	Table 9.1
Helper (age > 1 year), within natal territory	L_H	0.80 ($\male$)	Table 9.5
Breeder, overall average	l	0.82	Table 9.2
Breeder, without helper(s)	l_o	0.77	Table 9.4
Breeder, with helper(s)	l_h	0.85	Table 9.4
Fledgling production per individual (M)			
Average annual, all individuals lumped	$\overline{M}$	1.02	Tables 8.1, 8.2
Experienced breeder, with helper(s)[a]	M_{eh}	1.20	Table 8.4
Experienced breeder, without helper(s)[a]	M_e	0.82	Table 8.4
Novice breeder, with helper(s)[a]	M_{oh}	1.20	Table 8.4
Novice breeder, without helper(s)[a]	M_o	0.65	Table 8.4
Helper effect (average additional production)	ΔM	0.80	Tables 8.1, 8.2
Genetic relatedness (r)			
Age-specific, between helper (age t) and nestlings	r_t	$0.5(.82)^t$	Figure 5.2
Average, between mates	—	$0+$	Section 4.6[b]
Age at first reproduction[c]			
Percent of first-breeders, age:			
1 year	—	0.02($\male$), 0.05($\female$)	Table 4.1
2 years	—	0.39　　0.58	Table 4.1
3 years	—	0.33　　0.35	Table 4.1
4 years	—	0.14　　0.02	Table 4.1
5 years	—	0.06　　0.01	Table 4.1
6 years	—	0.04　　0.00	Table 4.1
7 years	—	0.02　　0.00	Table 4.1

TABLE 10.1. Continued

Description	Our Name for Variable	Observed Value in Florida Scrub Jay	Reference
Probability of obtaining breeding space (D,B)			
By dispersal, at earliest possible age[d]	D_o	0.42 (age 4 mo.)	Table 10.2
By dispersal, at age 1 year[d]	D_1	0.52	Table 10.2
By dispersal, annual rate for postyearling helper[d]	D_H	?	—
From within natal territory, average annual rate[e]	$\overline{B}$	0.40($\male$)	Table 7.3
From within natal territory, as sole helper	B_o	0.30($\male$)	Table 7.3
From within natal territory, as subordinate helper	B_s	0.30($\male$)	Table 7.3
From within natal territory, as dominant helper	B_d	0.56($\male$)	Table 7.3

Data for mixed pairs (novice/experienced) in table 8.4 are included by pooling their data, in appropriate proportions, with the respective categories here.
Close inbreeding is extremely rare; the actual figure for average relatedness between mates is imperceptibly greater than zero and has not been calculated.
Proportions listed from table 4.1 are those for first nestings, not first pairings.
See text and table 10.2 for methods of estimating D_o and D_1; values listed here are average estimates. D_H cannot be estimated for Florida Scrub Jays, because postyearling sample sizes do not include earlier-dispersing helpers. All estimates of D_o and D_1 assume perfect replacement of lost breeders, from a pool of potential recruits of equal age, and death of all additional, unsuccessful dispersers.
An independent estimate of $\overline{B}$ can be obtained by subtracting the annual death rate of male helpers at home (.14 to .22; see table 9.5) from their total average annual rate of disappearance as helpers, which is age independent (.54; see table 4.2 and figure 4.3). The result (.32 to .40) corroborates the value used here from table 7.3.

generation time, comparing the fitness of two traits requires analysis of the trade-offs between maximizing offspring production and adjusting the timing of this production (e.g., Southwood, 1981).

Remarkably, the literature with which we are familiar all but ignores the mathematically peculiar—but biologically common—special case in which the realized r is stable and exactly zero. In these cases, the fitness of individuals bearing a given trait becomes equivalent to their net production of breeding descendants (or the genetic equivalents produced by kin), *regardless* of how long these descendants take to become breeders (e.g., Horn, 1978). Generation time is irrelevant, at least in the simple analysis of whether a given trait can successfully invade a population bearing its alternative. (We ignore such complicated, and for present purposes irrelevant, problems as the genetic dynamics of a trait's spreading, and the resulting frequency-dependent changes in the selective advantages of the trait.) A similar observation is made by Vehrencamp (1979).

The Florida Scrub Jay presents a classic example of a stable population. Its habitat is sharply bounded and saturated, and the number of breeding individuals essentially does not change. Over the long term, the demographic stability within its population may be more absolute than in virtually any other bird. In such a regime individuals compete, not for some intrinsic rate of growth of their respective lineages, but rather for the total number of breeding descendants (or equivalents) *produced over the lifetime*.

Lifetime Reproductive Success

The concept of lifetime reproductive success and a strict accounting of all the demographic components that affect it are absolutely fundamental to our interpretation of cooperative breeding (see also Vehrencamp, 1979). To the extent that we differ with published discussions and models treating the evolution of helping, almost always the difference can be reduced to this issue. Before pursuing our own formulations, we illustrate this important point with examples drawn from three of the most important theoretical discussions of cooperative breeding.

1. *Kin selection*. Numerous theoreticians use the basic inequality of Hamilton (1964) to argue that helping behavior is *favored* by kin selection (italics ours). We have no quarrel with this postulate. It is a mathematical truism. To argue that helping behavior is *caused by* or *evolves through* kin selection is, however, a different matter. Brown (1975a; more precisely in 1978a) formulates Hamilton's conditions as follows:

$$\frac{G}{L_{max}} > R$$

where, in terms of our table 10.1, the following definitions apply:

$$R = \frac{r_{HO}}{r_{HR}}$$

$L_{max} = M' =$ young produced by unassisted pair.
$G = \Delta M =$ extra young produced by a helper.
$r_{HO} = .5 =$ relatedness to one's own offspring.
$r_{HR} = r_t =$ average relatedness to offspring being helped.

Brown shows that the above inequality rarely, if ever, is met in nature. He postulates—we think correctly—that the equation itself usually is erroneous. The variable L_{max} (referring to the potential gains in breeding alone rather than helping) must be corrected downward by the probability that an individual will be successful at obtaining a mate and breeding, should it choose to disperse rather than help. If, as we suspect, L frequently is close to zero for young individuals in crowded populations, then the above inequality does hold even where G or r_{HR} might be small. In Brown's words (1978a, p. 140), "The criterion for the individual to

313

help, $G/L > R$, is clearly met in such cases." The concern we have over this statement, and with most other references to Hamilton's inequality, is the implication that helping would not otherwise occur even if r_{HR} were zero or G included no increased production of siblings. To base theoretical conclusions upon the simple parameters G (measured in increased reproductive output) and r_{HR} ignores a host of *other* gains that might accrue to an individual by helping. Many of these other gains can be measured only over the total lifetime of the individual. Gains pertaining to survival, dispersal, territorial ownership, and later reproduction are not included in the simple inequality so often used to establish "the criterion for helping."

2. *Per capita fitness estimates in groups.* Koenig and Pitelka (1981) and Koenig (1981) provide a general, habitat-based model for the evolution of communal breeding. We fully endorse their framework: where habitat is rare, localized, and of relatively uniform quality (i.e., little marginally usable space), young individuals should remain in the natal territory or group (see below). However, in neither paper is the question raised as to why individuals who remain at home should bother *helping* in such an instance. Furthermore, the authors repeatedly refer to a measure, per capita fitness, the use and theoretical basis of which we cannot accept (see also section 8.9). These two shortcomings appear to result from an inadequate accounting of the lifetime rewards for the strategies in question.

Koenig and Pitelka (1981, p. 269) illustrate their habitat model with a useful analogy, the "post-doc" phenomenon: "The job market contracts; the optimal habitats fill; students 'post-doc' with their academic parent or with academic relatives and do so longer while waiting to disperse to a suitable site to independently enhance their professional fitness." The authors do not proceed with what we

314

consider the kernel of this analogy (and of this professional strategy): successful post-docs are not simply waiting in a queue and helping their established associates obtain better research results. Rather, they are *actively improving* their own chances to "disperse to a suitable site" by publishing papers, attending meetings, engaging in interviews (forms of dispersal forays), and performing other preprofessional activities that improve their average *lifetime* professional fitness. Post-docs, and helpers, can help themselves as long as the final analysis is not performed over the short run but over a lifetime.

Koenig and Pitelka correctly state (p. 266) that ". . . a theoretically straightforward test . . . is the lifetime reproductive success of grouped *versus* ungrouped individuals" (italics theirs). They proceed, here and elsewhere, to discuss the implications of a seemingly universal, negative correlation between group size and per capita reproductive success. Their approach is conservative, because helpers on the average are less related to offspring than are breeders, and they still correctly conclude that individuals always would be better off in pairs than in groups. In their treatments, however, per capita success always is calculated independent of the known or suspected genetic relatedness between group members, thereby rendering this index noncomparable between individuals (see section 8.9 for more discussion of this problem). Furthermore, the total benefits to the fitness of individuals cannot be derived from reproductive data alone, as Koenig and Pitelka derive it. Enhancement of survival probabilities (negated by Koenig, 1981, but demonstrated here in section 9.2) and improved opportunities for successful dispersal are two crucial benefits accruing to helpers but not involved in the short-term calculations produced from annual reproductive data alone.

Finally, it must be remembered that parents who allow

their offspring to remain home and post-doc safely can be improving their *own* reproductive success simply by improving the probability that these offspring eventually become breeders. Therefore, they might even accept a sacrifice in fledgling production and still come out ahead in overall lifetime fitness, whether or not per capita fledgling production is higher or lower. The reproductive value of older nonbreeding offspring (helpers) relative to younger ones (nestlings or fledglings) is a critical relationship that seems to have been ignored by virtually all treatments of cooperative breeding to date.

3. *Ecological constraints and behavioral conflict.* In two valuable papers, Emlen (1982a,b) lays out a conceptual model similar to that advanced by Koenig and Pitelka (1981) but more general. (Indeed, he points out that the habitat saturation model "has become the modus operandi for ecological thinking concerning the evolution of helping behavior" [1982a, p. 32].) Emlen (1982b) also models a conflict of fitness interests between parents and helpers, embracing the notion of Gaston (1978a) that in certain circumstances parents must receive "payment from their auxiliaries" in order to tolerate them. This circumstance arises when the reproductive costs to the parent in allowing their older offspring to stay home are not sufficiently compensated by increased reproductive output of these offspring once they become breeders.

An important feature is built into Emlen's models that we shall stress later, namely, the probability, ψ, of a young individual becoming established as a breeder. Emlen combines this crucial variable (called F by Brown, 1978a) with several others, into a single variable, W, his fitness parameter.

Emlen (1982b) builds his argument for breeder-helper conflict—as well as its resolution—around single-year val-

316

ues for W. Specifically, the crucial ratio is that between (1) the improved second year breeding output by the individual that helped its first year and (2) the absolute difference between annual parental breeding success with versus without helpers (op. cit., p. 43, equation 6). This formulation, while conceptually sound if generations are nonoverlapping, fails to include in the numerator the remaining lifetime gains by the helper. Adding these gains has the effect of reducing the potential area of conflict in Emlen's graphical representation. As life-span increases and generations become increasingly overlapping, the interests of the parents and offspring become increasingly similar. For organisms such as birds, the interests are very nearly identical.

We also suggest that the probability of becoming established as a breeder (ψ) be recognized as a variable quite separate from W. If probable fitness (i.e., total expected production of offspring) is measured over a lifetime, its value depends upon summed annual survival probabilities and fecundity values, all corrected by some probability less than one (ψ) that the individual will get a chance to breed at all. Including the variable, ψ, within W implies that it is a constant. We suggest that it can be varied and manipulated to a certain extent by the individuals, as discussed in the next section.

10.2. ALTERNATIVE STRATEGIES FOR NONBREEDERS

As indicated in the preceding section, we define the fitness of individuals pursuing a given strategy as the expected number of offspring or offspring equivalents they should produce during an average lifetime. In terms of direct reproduction alone, fitness (W) of an individual can be expressed as:

$$W = L_i D_i R \tag{1}$$

where L_i is the probability of surviving to dispersal age i; D_i is the probability of successfully becoming established as a breeder by dispersing at age i; $R = \sum_{i}^{n} l_x m_x$, or the total number of expected offspring from the first year of breeding onward (not to be confused with the demographic parameter, R_0, which already corrects for juvenile survivorship and any delays in onset of reproduction).

Individuals should delay breeding (with or without helping) when their fitness is enhanced by doing so. To formalize this truism, nondispersal should be favored when

$$W_h > W_o$$

where W_h and W_o represent the fitness of individuals following the helping or the early dispersal strategies, respectively. Substituting from equation (1):

$$W_h > W_o \text{ if and only if } L_h D_h R_h > L_o D_o R_o \tag{2}$$

This statement assumes (a) no indirect component to fitness and (b) identical survivorship during the time period between successful dispersal at ages h or o and the first breeding season. We can, of course, alter these assumptions. All subsequent discussions in this section proceed by modifying expression (2) and studying the results.

D_0, The Early Dispersal Quotient

All current notions about avian cooperative breeding include the observation that individuals who *can* disperse early to reproduce on their own *should* do so (e.g., Emlen, 1978, 1982a; Brown, 1978a; Koenig and Pitelka, 1981). This conclusion holds because the genetic gains accrued through production of sibs rarely approach those of direct reproduction, at least in birds (contra Brown, 1974, 1975a; Wil-

318

son, 1975; and Davies, 1982). For this reason, we find it useful to cast our fitness inequalities so that they explicitly relate to environmental conditions inherent in the value, D_o, the probability that early dispersers will become established successfully as breeders. As Emlen (1982a) points out, this parameter (related to his ψ variable, but specifically applying to prereproductive age individuals only) is affected by many features of the animal's physical, ecological, and demographic surroundings. It is a complicated variable, virtually impossible to measure objectively in the field. In the Florida Scrub Jay it can be viewed operationally as the quotient between the predicted number of successfully paired and breeding dispersers (measured at age 1 year) and the total number of independent young that would have entered the competition in the preceding autumn. (It should be directly measurable for Scrub Jays in western North America, because all young do disperse in their first autumn, and virtually all apparently are breeders the following spring.)

Because we view this quotient as so important, we set up our inequalities below as if to ask one simple question: at what critical value of D_o does the optimal strategy switch from early dispersal to its alternative? We examine several such alternatives and show where the most important variables contribute to the solutions to this key question. By focusing on this one variable, we risk disguising other equally critical ones inadvertently. Our intent is merely to draw attention to the dispersal parameter, a variable that has received little formal attention to date in the literature on cooperative breeding.

Yearling versus Juvenile Dispersal

We begin with the simplest choice, one between dispersing at the earliest possible age (e.g., 4 months for Scrub Jays in western North America) versus delaying dispersal through age 1 year and then dispersing to begin breeding at age 2. We can simplify the expression for conditions favoring the strategy by defining a new survivorship measure, L_1, to be the proportional survivorship of nondispersing individuals between earliest possible dispersal age and age 1 year (i.e., the ratio between L_h and L_o). This ratio will always be less than one. Secondly, if we assume that the total survivorship and reproductive output as a breeder is equal between the two strategies, then the conditions for delayed dispersal become:

Define: $\quad \dfrac{L_h}{L_o} = L_1 \quad$ and $0 < L < 1$.

If $\qquad\qquad L_h D_1 R_1 > L_o D_o R_o \qquad$ from (2)

assume: $\qquad\qquad R_1 = R_o$,

then $\qquad\qquad D_o < L_1 D_1$ $\qquad\qquad$ (3)

Expression (3) is the intuitively obvious statement that delayed dispersal is favored whenever its likelihood of success is sufficiently greater than that of early dispersal to offset the probability of dying before dispersing, assuming that the latter is the only cost to the delay.

Yearling Helper versus Juvenile Disperser

Now suppose that the delayed disperser helps at home, resulting in some indirect benefit, k, through the increased production of relatives. In most avian cooperative breeders, the helper assists its parents or stepparents, whose annual survivorship is l. The increased (or decreased) production of sibs ($r = .5$) because of the helpers is represented by M (see table 10.1). In such a case, the value of k in offspring equivalents is $(l)^t \cdot \Delta M$ (Brown, 1978a). In the present example, $t = 1.0$ because the single year's delay is followed by dispersal. In the Florida Scrub Jay, therefore, $k \approx .66$ (table 10.1). With this added benefit to remaining home, and maintaining the assumptions of expression (3), expression (2) must be modified to the following:

$$L_I(D_I R_I + k) > D_o R_o \qquad (2')$$

or,

$$D_o < L_I(D_I + \tfrac{k}{R}) \qquad (4)$$

The conditions favoring helping thus depend principally upon three terms: (1) the ratio between indirect (k) and the direct (R) expected offspring production; (2) the chance that a yearling will successfully disperse; and (3) the chance that an independent juvenile will survive at home to become a helper. If dispersal probability is approximately age independent ($D_o \approx D_I$), expression (4) becomes:

$$D < \frac{k L_I}{R(1 - L_I)} \qquad (5)$$

Inequality (5) represents a simple kin-selection model for the evolution of helping behavior, graphically represented in figure 10.1. In words, it states that helping by yearlings should evolve when the risk of death through early dispersal is greater than the expected indirect benefits that can be gained through helping. The magnitude of the overall benefit is proportional to the *relative* magnitude of the indirect component (compared to direct reproduction) and to the survival probability of a young individual remaining at home (dashed lines in figure 10.1). As juvenile survivorship in the natal territory increases, it becomes increasingly profitable to remain at home and reap the indirect benefits (k) of helping. Put another way, for any fixed dispersal probability (D'), the magnitude of the indirect benefits required to favor helping decreases as juvenile survivorship *at home* increases. The only cost to staying at home is the $(1 - L_I)$ probability of death before breeding.

The relationships within expression (5) and in figure 10.1, and the assumptions included therein, illustrate com-

321

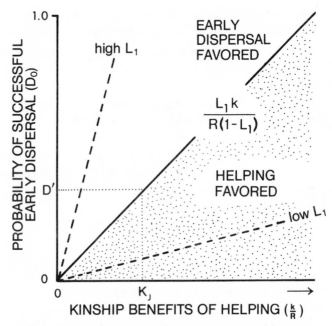

FIGURE 10.1. Simple kin-selection model for evolution of helping behavior. Juveniles in population (J) should stay home and help if their chances of successfully dispersing (D) are less than D', because they can count on sufficient offspring equivalents being gained through helping. If $D_J > D'$, their expected fitness is greater by risking dispersal than by risking death at home before breeding. Slope of critical D function is determined by juvenile survival (L_1), as shown by dashed lines. See text for assumptions.

mon features of kin-selection models in general. Note, for example, that if k were zero (or even very small relative to R), helping would not arise because nothing is to be gained, and early dispersal would always have at least some finite probability of success. This conclusion is an explicit example of a concept we find implicit in much of the sociobiological literature, namely, the notion that without an indirect kinship component to inclusive fitness, helping behavior would not arise. Conversely, if opportunities to disperse successfully early in life are abundant (D_o close to 1.0), unrealistically enormous indirect benefits (relative to direct ones) would be required to favor a helping strategy. The fact that this environmental condition is rare among cooperative breeders leads to the nearly universal conclusions (references cited above) about the pervasive importance of severe ecological constraints in the evolution of helping. Unfortunately, it also makes difficult the task of separating the relative contributions of the indirect and direct components.

To review our assumptions thus far, expression (5) assumes: (a) dispersal capabilities are equal for independent juveniles and yearlings (i.e., $D_o \approx D_l$); (b) helpers remain for only one year, then disperse and either breed or die; (c) as breeders, individuals show similar l_x and m_x values (survival and fecundity) regardless of the strategy they followed to become breeders.

A more realistic scenario would involve some biological difference between the dispersal capabilities of individuals at the earliest possible dispersal age and at age 1 (probabilities D_o and D_l in expression [4]). The likeliest situation—undoubtedly characterizing Florida Scrub Jays—is that older individuals would have greater probability of succeeding. (This need not be the case, however; in some situations, such as overwintering as prereproductives, dispersal probability may decline with age, producing equally interesting predictions about life history tactics [see Horn, 1978].) We can define a new value, I, as the ratio between later versus earlier dispersal probabilities, D_l/D_o.

In Scrub Jays and other cooperative breeders, I almost always is greater than 1.0. Expression (4) becomes:

$$D_o < L_I D_I + \frac{L_I k}{R} \qquad \text{from (4)}$$

$$\text{or,} \qquad 1 - L_I \frac{D_I}{D_o} < \frac{L_I k}{R D_o}$$

$$\text{Define: } \frac{D_I}{D_o} = I; \qquad 1 - L_I I < \frac{L_I k}{R D_o}$$

$$\text{or,} \qquad (1 - L_I I) D_o < \frac{L_I k}{R} \qquad (6)$$

The form of expression (6) is identical with that of (5). The only difference is that the cost of staying at home is reduced (increased if $I < 1.0$) by the inclusion of I in this term. The term appears in the left-hand side to avoid division by zero, as the expression $(1 - L_I I)$ can be either positive, zero, or negative. If it is negative (sufficiently large I), the helping strategy is favored under all possible permutations of L_I, k, D_o, and R, because the right-hand side is always either positive or zero. This is an extremely important result: in the previous model (figure 10.1), helping was favored only by the existence of a positive indirect benefit. It would represent a pure form of kin selection. In the present case (figure 10.2) helping (or at least delayed dispersal) can be favored *even in the absence of any kinship benefits whatsoever* (in fact, k can be negative and still produce a stay-at-home strategy). The only criterion is that dispersal probabilities following the delay be sufficiently enhanced compared to early dispersal.

The above decoupling of kinship benefits and dispersal benefits inherent in helping was proposed by us much earlier in a verbal model that was specific to a few features of Florida Scrub Jay sociality (Woolfenden and Fitzpatrick, 1978a). It represents our major departure from typical kin-selection models (West Eberhard, 1975; Wilson, 1975;

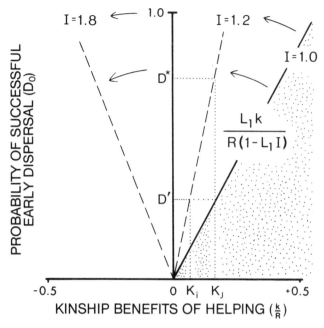

FIGURE 10.2. Dispersal-mediated modification of model in figure 10.1. Given an early-dispersal probability of D', individuals receiving less than K_j benefits from helping should disperse in fig. 10.1; here, if their dispersal probability as yearlings is 20 percent higher ($I = 1.2$), they should still stay home and help. In this instance, only if their indirect benefits are below K_i should they disperse early. For individuals in population J, a 20 percent increase in dispersal capability by remaining one year demands a nearly three-fold increase in early-dispersal probability (D' to D^*) before early dispersal is favored. If later dispersal is sufficiently more probable than early dispersal (e.g., $I = 1.8$), individual should stay home regardless of any presence or absence of kinship benefits there.

Brown, 1974, 1975a, 1978a; Vehrencamp, 1979). Once decoupled into separate terms, the relative values necessary to favor one strategy over the other can be compared, both hypothetically and using real data. Natural systems can be examined and compared with respect to how important these two parameters (among others) are relative to each other. Before we do this here, however, we propose one further modification that more closely approximates many avian communal breeding systems, including that of the Florida Scrub Jay.

Stay-Home-and-Wait versus Early Dispersal

Until now we have been assuming some constant probability of successful dispersal at age 1 year. If this probability is sufficiently higher than that for early dispersal (for any of a variety of reasons), delay is favored. Now suppose that the nondispersal strategy is extended indefinitely. Instead of leaving home at age 1 (perhaps after helping) the prebreeder simply remains at home until a breeding opportunity presents itself. For the present, the nature of these opportunities is unimportant. The critical point is that physical risks in the sometimes prolonged act of dispersing are greatly reduced, perhaps even to a negligible level. However, two new variables now enter the picture. One (B) is simply the annual probability that an opportunity will indeed arise for a given individual. The other (L_H) is the annual survivorship of the helper who is staying at home. B is affected by external features of the demographic regime (much as D_1 and D_o are) and by any number of behavioral strategies that might be employed from within the territory (e.g., helping, cooperation, dominance, territorial expansion, inheritance). L_H also is under some external control (through predator pressure, incidence of disease, food limitation, etc.), but presumably will always be relatively high compared to L_1 (when the juvenile still is gaining experience). Helper survivorship at home even may be expected to exceed breeder survivorship (chapter nine).

We presented evidence in chapter seven that Florida Scrub Jay helpers behave as if this "passive" strategy is being followed (especially note figure 4.3). With relatively constant annual values of B and L_H, departures from home through several years of helpers' lifetimes would be expected to exhibit a negative exponential curve. Indeed, they do, at least in the case of males.

How does this change in strategy affect our model (besides making it more complicated to follow)? We return to expression (2') and expand it:

$$D_o R_o < L_l D_l R_l + L_l k.$$

Assume: $\qquad R_o = R_l, \text{ and } D_l = \psi:$

$$D_o < L_l \psi + \frac{L_l k}{R} \qquad (7)$$

The variable D_l must be replaced by an expression that represents the total, summed probability of becoming a breeder by remaining at home. We use ψ, recognizing its similarity to the analogous variable in Emlen's (1982a) model. However, our ψ refers to an overall postyearling dispersal probability for individuals remaining at home to yearling age. To assess its real-world characteristics, we must express ψ in terms of L_H and B. Specifically, it can be shown that:

$$\psi = B \cdot \sum_{i=1}^{n} (L_H)^i \cdot (1 - B)^{i-1} \qquad (8)$$

That is, ψ is the product of the annual probability of becoming a breeder and the summed probability of survival through the years until a vacancy arises.

Expression (7) continues to assume that the total reproductive output after breeding status is achieved is not affected by the dispersal strategy used. It also includes only one year's indirect benefit, accrued through helping to raise sibs. Although it can be modified easily, we maintain this assumption for two reasons. First, average genetic relat-

327

edness between helper and offspring declines with age, owing to parental mortality (as also shown by Brown, 1978a). Second, and more important, each year's potential indirect benefit, k, must be corrected for the probability that a surviving, younger sib would be helping in the same territory, thereby producing the extra offspring even in the absence of the older, potentially dispersing helper. (If parental reproduction did increase with the number of older offspring helping, this correction factor would be smaller and considerably more complicated; however, in the Florida Scrub Jay and in all cases reviewed by Koenig, 1981, a solitary helper contributes essentially as much as any nonzero number of helpers, so our assumption is valid at least in these avian examples.)

We substitute expression (8) into (7):

$$D_o < \frac{L_I k}{R} + L_I B \cdot \sum_{i=1}^{n} (L_H)^i (1 - B)^{i-1} \tag{9}$$

Solutions to expression (9) are graphed in their general form in figure 10.3, and with actual survivorship and kinship values in figure 10.4. Characteristics of this general model, including its implications regarding kin selection, are discussed in the next section.

10.3. BREEDING-SPACE COMPETITION AND EVOLUTION OF COOPERATIVE BREEDING

The conditions under which helping should arise in a seasonally reproducing, stable, territorial bird population principally depend upon the relationships between eight demographic variables, all of which are included in expression (9), above. For review, these are: (1) the probability that an individual, after reaching the earliest dispersal age, can at that time successfully disperse and become established as a breeder before the following breeding season (D_o); (2) among nondispersing individuals, the survival probability in the natal territory, between early dispersal age and age 1 year (L_1); (3) the average number of offspring equivalents produced by an assisted pair, beyond what they would normally produce in the absence of help (ΔM); (4) the average relatedness between the helper and these "extra" offspring ($= r_i$; variables ΔM and r_t together determine the value of k in expression [9]); (5) survivorship as a breeding adult (l_x); (6) average annual production of offspring ($= M$; variables l_x and M together determine the value of R in expression [9]); (7) survivorship as an older, experienced helper before becoming a breeder (L_H); and (8) the annual probability that a surviving helper, remaining in the natal territory, will encounter (or create) a breeding vacancy it can successfully fill (B).

More than eight variables actually are involved, of course, especially in the determination of R (net lifetime reproductive potential), given varying degrees of experience and helper effects on reproduction and survival (table 10.1 lists many components to R). For purposes here, however, we can visualize more easily the form of our model by lumping these, plus kinship values, into a single expression, $\frac{k}{R}$, which we might call the "kinship coefficient." This ratio assesses the relative magnitude of any indirect kinship component to fitness, compared to all possible direct components. As such, it is closely related to Vehrencamp's (1979) "index of kin selection," but is calculated quite differently. (We stress, for example, that average relatedness to recipients of help declines as the helper gets older; this observation was overlooked by Vehrencamp.) Like our B, the kinship coefficient is implicitly involved in all discussions of the role of kin selection in avian cooperative breeding, although it has sur-

329

faced explicitly only rarely (notably in Veherencamp's [1979] excellent review).

That the ratio, $\frac{k}{R}$, is a critical measure of the role of kin selection in any system, we find intuitively obvious. If k is zero (no kinship benefits to helping), kin selection cannot occur, by definition. If k and R are roughly equal (ratio near unity), clearly the role of the indirect component to fitness is relatively enormous. The role of kinship becomes obvious where k is much greater than R (as among worker castes in some social Hymenoptera).

The conclusions we draw from our formulations rest primarily on the observation that a nondispersal strategy involving helping behavior can evolve as a result of breeding space limitation, in the presence *or the absence* of kinship benefits associated with this strategy. While kin selection can in certain cases cause a strategic shift from early dispersal to helping, the exact same shift can arise from conditions that are independent of kinship. We make five specific points along these lines, each of which is illustated in figures 10.3 and 10.4.

1. In the system we have constructed, kinship components to inclusive fitness clearly provide conditions under which helping would otherwise not evolve. In the general case shown by figure 10.3, any combination of dispersal values falling below the solid line will favor delayed dispersal. Below the dashed line ($k = 0$), a delay is favored even in the absence of any kinship benefits that might accrue through helping. In the zone between the lines, however, delayed dispersal is favored only because the average helper raises extra kin, thereby adding sufficiently to inclusive fitness to offset the risk of dying before dispersing. The values singled out in figure 10.3 illustrate a case in point. In the absence of kinship benefits in a population where, by delaying dispersal, individuals realize a proba-

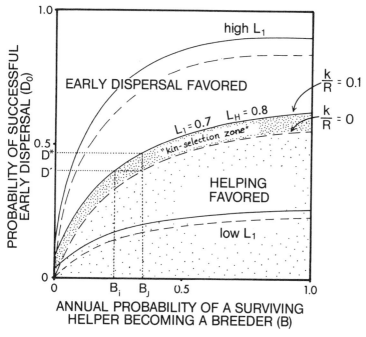

FIGURE 10.3. General configuration of the solutions to expression (9) in text. For low, medium, and high values of juvenile survivorship (L_1) dashed line shows critical D_o function where k/R (kinship coefficient) is zero, solid line where $k/R = .1$. In populations where individuals' average D_o and B values fall under the critical D_o function, delayed dispersal and helping are favored. Lightly stippled area shows values where delayed dispersal and helping would be favored in the absence of a kinship component to inclusive fitness; heavily stippled area ("kin-selection zone") shows values where kinship benefits cause helping to evolve where it otherwise would not.

bility less than B_j of becoming breeders each year, early dispersal would be favored given D' or better chance of its succeeding. Now suppose that additional sibling production in this case provides a finite gain in offspring equivalents ($\frac{k}{R} = .1$ in the example shown). In this case the individual could tolerate an annual breeding probability as low as B_i (or, forgo an early dispersal probability as high as D^*) and still realize greater lifetime fitness by helping rather than dispersing early. In figure 10.3, we refer to this demographic area, in which helping is favored only because of the existence of kinship benefits, as the "kin-selection zone."

2. Delayed breeding and helping can evolve in the absence of kinship benefits when the combination of dispersal probabilities thereby realized places the population below the dashed line in figure 10.3. For populations with these characteristics helping behavior can be favored or reinforced by kinship, but this need not be involved in the evolution of the system. Indeed, a negative kinship coefficient (resulting from "hindering-at-the-nest") could be involved (solid line falling below the dashed one in figure 10.3) while helping still is favored. The criterion, at least from the potential helper's point of view, is that the system produce sufficiently higher annual probability of locating a breeding vacancy to offset the risk of dying before leaving home.

3. The existence of a favorable kinship coefficient (positive $\frac{k}{R}$) always relaxes the conditions under which helping should arise. However, only at relatively high values ($\frac{k}{R}$ between .2 and 1.0) is this range of conditions increased significantly (figure 10.4). Especially when prereproductive survivorship values are relatively high (as in most cooperative-breeding birds), a kinship coefficient between 0 and .1 does not substantially alter the shape of, or area under, the fitness curves. At high values of $\frac{k}{R}$, such as those among some social insects, cooperative behavior is predicted under any but the most lenient possible dispersal regimes. Incidentally, this means that an order-of-magnitude estimate of the kinship coefficient for a given population provides at least a rough guess as to the probable importance of kinship in the evolution of its social system. Among co-

operative-breeding birds, kinship coefficients appear to vary from .02 to .1 (calculated from the equation $k = (l)^l \cdot \Delta M$, defined above, and using the data in Brown, 1978a, and Koenig, 1981).

4. Delayed breeding and helping can be favored even under conditions where the probability of success at early dispersal (D_o) is *higher* than the subsequent annual probability of locating a breeding space by staying home. The dashed lines in figures 10.4A,B,C show points of equality between D_o and B. Under most combinations of values, a substantial zone exists in which helping results in higher fitness even though early dispersal probability is higher than B (e.g., shaded area in figure 10.4A). In demographic environments of low nonbreeder survivorship (figure 10.4B), substantial kinship benefits are required for this to be the case. Where nonbreeder survival is high, however, a large area exists above this line of equality even where kinship benefits are low or nonexistent (figure 10.4C). This occurs in cases of relatively low expected success at dispersal and low annual appearance of breeding vacancies, combined with high survivorship at home—precisely the conditions characterizing the demographic environment of the Florida Scrub Jay.

5. Because the fitness curve is highly nonlinear at low values of B, any impact that helping behavior per se might have on B will be especially important in this zone. Put differently, in environments where the annual probability of obtaining breeding space is very low, any trait or activity through which an individual can raise this probability, even slightly, will substantially increase the likelihood that delayed departure from home will be favored over early dispersal. This point becomes increasingly important as helper survivorship in the home territory (L_H) increases, as shown

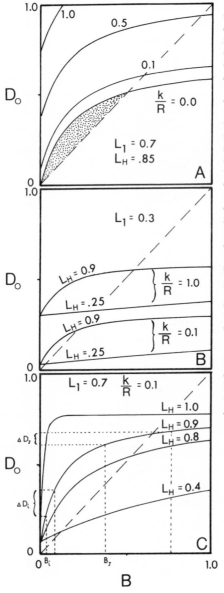

FIGURE 10.4. Sample solutions to the inequality in expression (9) in text, using different values for L_1 (juvenile survivorship), L_H (helper survivorship), and k/R (kinship coefficient). Probabilities of successful early dispersal (D_o) are plotted against probabilities of surviving helper obtaining a breeding space by staying home and helping (B); at each combination of values, the helping strategy results in greater fitness below the line, while early dispersal is favored above it. Each function's Y-intercept is determined by k/R (e.g., fig. 10.4A) and L_1 (e.g., fig. 10.4B), while its slope and degree of nonlinearity are most affected by L_H (e.g., fig. 10.4C). Diagonal dashed lines show values where D_o and B are equal. As shown in fig. 10.4C, doubling the annual breeding probability has much more influence on the critical D_o value at low values of B (e.g., $B_i, \Delta D_i$) than at higher ones (e.g., $B_j, \Delta D_j$).

in figure 10.4C. Kinship does not affect the nonlinearity of the relationship; hence it is irrelevant to this point.

Conclusions from the Model

Delaying dispersal until a breeding vacancy can be located, or created, from within the natal territory forms the substance of the strategy we are analyzing. Conditions under which this strategy should be favored over early and permanent dispersal depend upon the precise relationships between many demographic variables. When juvenile and helper survival within the natal territory is high, the stay-home (or "work-from-within") strategy is favored when restricted breeding space creates a low probability of successful early dispersal. Among populations where the annual availability of breeding vacancies is low (i.e., in saturated habitats with long-lived breeding adults), behavioral strategies that improve nonbreeder survival and that raise, even slightly, the probability of obtaining breeding space will be favored, and will substantially reduce the conditions under which early dispersal should be favored. Group living, cooperative territorial defense, dominance, territorial expansion, and inheritance represent examples of such strategies. Kin selection also relaxes the conditions under which early dispersal is favored over the helping strategy. The overall importance of this reduction depends upon the ratio between the offspring-equivalents that can be added to the population indirectly and the total number of direct offspring that can be expected once independent breeding status can be attained. In most birds, this ratio appears to be small (less than .1), especially in comparison to the range of values that are possible among such animals as haplo-diploid insects.

Demographic Attributes among Scrub Jays

Where do Florida Scrub Jays fall within the picture just painted? Just as interesting, what can we predict about the population biology of Scrub Jays in western North America, where the helping strategy apparently is not favored? Figure 10.5 summarizes our answers to these two questions. Much of this book was devoted to establishing values with which to answer the first question. Unfortunately, we can only make general statements regarding the second.

Using the data from table 10.1, we have constructed figure 10.5 to summarize the Florida Scrub Jay demographic picture relative to our concepts about the evolution of helping. The curve is drawn using a kinship coefficient of .1 (k = .66, R = 5.25; actual value = .12; values calculated from data in tables 9.10 and 10.1), and for perspective we illustrate the curve for k/R = 0. The juvenile survivorship value, L_I, is obtained by dividing the survivorship value for one year olds (L_h) by that for independent juveniles (L_o). Helper survivorship in the natal territory (L_H) is taken from table 10.1.

Shown under the curve in figure 10.5 is the range of values for the two dispersal variables of interest, stippled to indicate the approximate mean values, in our Florida Scrub Jay population. Our estimate for D_o is indirect, as we have witnessed too few adventurous early dispersers to calculate directly their probability of success in our population. Table 10.2 presents the data from which the range and mean values were taken. Our estimates of B (table 10.1) ultimately derive from the rate at which helpers are observed to become breeders (see especially figure 4.3), and the range is obtained from the values for dominant versus subordinate helpers (table 7.3).

Clearly, the range of values now characterizing the Flor-

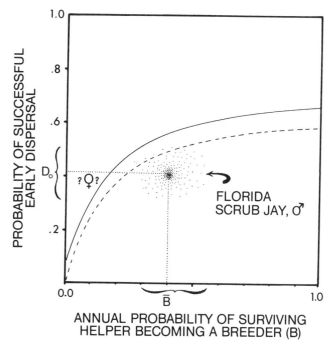

FIGURE 10.5. Approximate positions of Florida Scrub Jay males (stippled region) and females (with question marks) in relation to the actual values of the critical D_o function in the study population ($L_1 = .7; L_H = .85, k/R \approx .1$). The helping strategy is clearly favored, among males at least, even if kinship benefits were lacking ($k/R = 0$, dashed line). We cannot measure females' B-values directly, because they begin actively dispersing at, or shortly after, age 1 year. Because females rarely inherit breeding space even when they remain home, their annual probability of becoming a breeder by staying home is much lower than in males. Perhaps this low value places females *above* the critical D_o function, thereby explaining their much earlier departure from the natal territory.

ida Scrub Jay demographic regime places our population well within the zone at which delayed breeding and helping are advantageous over early dispersal. Comparing figure 10.5 with figures 10.3 and 10.4 shows that the relatively high values for juvenile survivorship at home contribute

Table 10.2. Estimation of Average Probability of Successful Dispersal at Age 4 Months (D_o), Assuming Exact Replacement of Lost Breeders by Dispersing Juveniles

| | Breeding Season | | | | | | | | | Three Estimates of D_o | | |
	1970	1971	1972	1973	1974	1975	1976	1977	1978	From pooled sample	Mean annual	Direct calculation
Breeders[a]	34	52	65	55	57	55	61	62	68	509	—	—
Breeder mortality	.118	.308	.154	.073	.246	.127	.262	.065	.265	.180	—	.180
Breeding vacancies	4	16	10	4	14	7	16	4	18	93	—	—
4-mo. juveniles[b]	22	19	16	19	18	36	17	35	38	220	—	—
Average D_o	0.18	0.84	0.63	0.21	0.78	0.19	0.94	0.11	0.47	0.42	0.48 ± 0.3	0.37

NOTE: Calculation of D_o directly from demographic averages: (breeder death rate) ÷ (fledgling production) (4-mo. survival) = (.180) ÷ (1.02) (.48) = .37; data from tables 9.2, 8.1, and 9.1, respectively. Analogous calculation for yearlings (D_1), using 12-month survival data: (.180) ÷ (1.02) (.34) = .52.
[a] Data from table 9.2.
[b] Data from table 9.1.

substantially to this result. We note that the Florida Scrub Jay falls well below the kin-selection zone, from which we conclude that kin selection has not played an integral role in the evolution of helping in this system. The system appears to be strongly favored by direct, cost-benefit relationships regarding breeding space competition, quite independent of the addition of a kinship component to inclusive fitness.

The Scrub Jay in western North America follows an early dispersal strategy. From this, we can only predict that its demographic picture would include a range of dispersal values that place its populations *above* the fitness curves as drawn for those populations. This could result from a number of independent differences with respect to the Florida population. Their overall survivorship values as nonbreeders—even in the natal territory—could be lower than in Florida. More likely, their opportunities to disperse successfully as juveniles could be substantially higher. Using figure 10.5 as a model, for example, if independent juvenile Scrub Jays in western habitats can survive on their own (or in juvenile flocks) at about the same rate as in Florida ($L_H = .7$), and if essentially all surviving juveniles can breed at age 1 year, then their D_o is also about .70. At this value figure 10.5 shows that a young jay should always disperse.

We predict that a fundamentally different value or combination of values characterizes the demography of the western Scrub Jay populations, and that the critical difference probably arises from the relatively broad habitat opportunities of those populations. A wide range of acceptable habitats means that dispersing juveniles need only to stay alive in order to be more or less certain of obtaining space in which to breed as yearlings. This simple explanation for the absence of helping behavior in western North American

Scrub Jays now needs to be tested against some actual data for those populations, data that remain unavailable to date.

10.4. THE EVOLUTION OF COOPERATIVE BREEDING IN *APHELOCOMA*

The genus *Aphelocoma* consists of three species. The social organization of one, the Unicolored Jay, is virtually unknown except that it appears to breed in groups (Pitelka, 1951; Goodwin, 1976; Brown, personal communication). Information on the social structures of the other two (Scrub Jay and Mexican Jay) is extensive. Based upon this latter information, we propose here a historical sequence—from strictly monogamous breeding to cooperative breeding— as it might develop in a bird such as a jay. In doing this, we still do not assume that cooperative breeding evolved on its own within the genus (see chapter three). Cooperative breeding well may be the primitive condition for *Aphelocoma*, and through time Scrub Jays in western North America could have lost this ancestral trait. However, we consider it useful to contemplate the origin of this rather rare and complex behavior from a common, widespread, and relatively simple social system such as breeding in unassisted pairs. The scenario we develop follows the demographic logic of the preceding sections and describes a series of changes in social organization ultimately brought about by changes in habitat (figure 10.6; see p. 343).

Southwestern North America represents a large and diverse segment of a once much larger area of xeric habitats that included a variety of dry, sclerophyll woodlands (Axelrod, 1958). Rainfall and temperature presumably varied considerably across this ancient, Madro-Tertiary habitat. Within it, the proto-Scrub Jay maintained a social organization much like that of the modern populations on the

North American mainland. Early, permanent dispersal was possible because the extent and variety of suitable habitats included marginal and medium-quality habitats as well as optimal ones for breeding. Annual variations in environmental factors prevented the jay population from thriving continuously at saturation levels. Much the same situation still characterizes today's western Scrub Jay populations.

Through a rise in sea level or dispersal, a population of Scrub Jays became isolated on Santa Cruz Island, in the Channel Islands off coastal California. Here, optimal and marginal habitat still exist, but limited area combined with a maritime climatic regime reduce the population fluctuations. The optimal habitat appears to be saturated with breeding pairs. The existence of suboptimal breeding habitat allows the persistence of early dispersal, but limitations in the available area have resulted in delayed breeding among the dispersers. As a result, on Santa Cruz Island troops of nonbreeding Scrub Jays include individuals several years old. They roam through suboptimal habitat, constantly testing the territorial occupancy of pairs in the better habitats. When a vacancy or weakness appears, members of the roving flock vie aggressively to fill in the position (Atwood, 1978, 1980a,b).

We suspect that the explanation for the social differences between the Santa Cruz and the Florida populations of Scrub Jays lies in the availability and demographic conditions of a habitat that can support nonbreeders outside of the main breeding habitat. Our demographic model predicts that Santa Cruz juveniles, while living and wandering through suboptimal habitat, experience higher survival than would analogous, juvenile dispersers in Florida. We predict, in addition, that on Santa Cruz Island the dispersers encounter openings in the breeding population that they can fill at age 1, at least occasionally. (The same demo-

FIGURE 10.6. A schematic pathway of social evolution in the corvid genus *Aphelocoma*. Nestings are represented by the letter *N* below mated pairs. Helping at the nest is represented by dotted lines between helper and nest. In the socially primitive condition (1) offspring depart the natal territory upon nutritional independence. In the Scrub Jay of western North America, these juveniles join nonbreeding flocks and breed at age 1 year. High dispersal-related mortality early in life, amidst a hostile or crowded environment, selects for delayed departure from home (2). This in turn leads to delayed breeding and brief bouts of helping at the nest before departure (3). As dominance and inheritance come into play (4), helping is reinforced; helpers, especially males, obtain immigrant mates, bud off a new territory from expanded natal ground, and breed there, defending the ground from the natal family with the help of the mate. This stage (4) characterizes Florida Scrub Jays. In the most critical stage for advanced sociality (5), the defense of a firm boundary within a clan diminishes, and older helpers help at more than one nest. This may result from increased longevity of the helpers, causing sibling partnerships to last into breeding (see text), or from an increased frequency of joint dispersals by females who then pair within one territory. Finally (6) the territory becomes entirely jointly defended, and both nonbreeders and breeders help at all nests. Immigrant mates do not attempt to defend their own territory, but simply join the clan. This system characterizes the Mexican Jay.

graphic result would obtain if breeding were sometimes possible in the marginal habitat, resulting in sufficiently high probability of early breeding.) This *combination* of demographic parameters—higher juvenile survival, lower breeder survival, and probability of breeding at age 1 year—raises the overall *D* for a Santa Cruz jay high enough to select for early dispersal, even though dispersal requires a transition period spent in marginal habitat. Of course, the scenario just developed is a simplified one, in that other variables (especially net reproductive output with versus without helpers and survivorship in groups of various sizes) also are involved, as modeled above.

Dry, sclerophyll woodland is a relict habitat in Florida. It occurs in scattered patches, many of them no more than a few square kilometers in area. In peninsular Florida, prairies, pine flatwoods, and hardwood hammocks now dom-

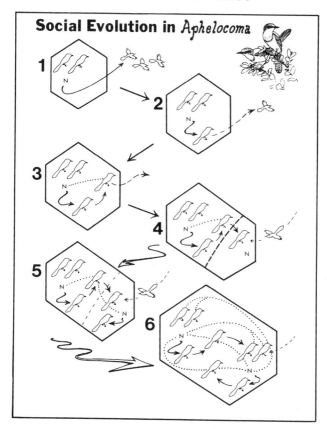

inate. These habitats are not inhabited by Scrub Jays, even wandering ones.

As moist environments spread through Florida during the early Pleistocene, Scrub Jays were cut off from their western relatives. Remnant populations were faced with living in smaller and smaller patches of relatively uniform scrub. Blue Jays may have prevented their spread into adjacent woodland habitats. Increasingly, patches of scrub became saturated with territorial breeders, leaving no convenient habitat in which young dispersers could survive. As the system evolved, early dispersal began to result in death.

The first major alteration of the Scrub Jay social system, leading later to cooperative breeding, was that breeding adults began to allow their maturing young to stay at home well beyond nutritional independence. The step simply represents an extension of parental care in a crowded environment. Such a satellite existence of young jays about their parental territory is reminiscent of the behavior of young Gray Jays, *Perisoreus canadensis* (Rutter, 1969) and Blue Jays (e.g., Hardy, 1961; J. Cox, personal communication) at some localities in their extensive ranges.

The next step in our postulated development of cooperative breeding literally defines the phenomenon. This, of course, is the act of helping by jays that have remained at home through the onset of a new breeding cycle. As with the act of remaining home in the first place, breeders presumably can prevent the acts of helping but cannot enforce that they be performed (contra Alexander, 1974). Why should breeders allow these nest visits, and why should the nondispersing young bother to perform them?

We have shown that pairs with helpers clearly produce more young, and they also show reduced mortality. Alone, these direct benefits could cause the breeders to allow helping at the nest. If helpers themselves also live longer, or enjoy improved performance as breeders because of helping, then again the original parents benefit directly in improved Darwinian fitness. If breeders can actually help their offspring to gain breeding space, then once again the breeders gain directly. These manifold advantages all derive from the extension of parental care to individuals older than one year, whose reproductive value far exceeds that of the nestlings being helped.

Why the helpers bother to help is perhaps the most critical question pertaining to cooperative breeding. The direct and indirect benefits that result from helping to feed young at the nest have been described and discussed throughout this chapter and need only be summarized here. Active helping might initially have resulted in larger family sizes, thereby increasing survivorship of the helpers (direct benefit) and their parents (indirect benefit), and increasing the representation of siblings in future generations (indirect benefit). Larger families presumably enjoyed advantages in the competition for breeding space (direct benefit), through increased territory sizes, inheritance of natal ground, and a decreased need to disperse from the territory in order to locate breeding vacancies. Perhaps the act of delivering food to nestlings reinforced a dominance hierarchy within the family, which would become important when siblings of varying ages began to coexist in the same groups (direct benefit). As longevity of adult and young jays increased, and as available habitat continued to decrease in Florida, helping at the nest became fixed in the population.

To the above capsule of benefits associated with helping behavior, we must add a caveat. We believe it quite possible that the act of delivering food to dependent young actually is selectively neutral. Feeding young at a nest in fact may be of no energetic consequence to a helper, and instead may represent a hormonallly controlled response to the presence of a nest containing squealing young jays. The classic example of nonadaptive feeding between Northern Cardinal and goldfish (Welty, 1975), as well as similar cases, indicates that the drive to feed young can cause nonparental feeding among birds. Furthermore, Florida Scrub Jays often vigorously defend nests that have been empty

345

for several days, either through predation or fledging. These nests never are used a second time, yet the birds clearly retain a strong tendency to defend them despite the absence of the young. Only circuitous reasoning allows one to supply an adaptive scenario for the so-called evolution of this behavior. More likely, we think, this behavior is a behavioral response with some hormonal basis, bearing no adaptive value and waning as the normal stimulus for it fades into the past. We cannot help but draw analogies between this nesting response and the act of feeding young by helpers who find themselves associating with a nest full of helpless and hungry young jays.

With only slight shifts in the demographic picture, the single-pair territorial system typical of Florida Scrub Jays easily could become the multipair system of the Mexican Jay. Our examples of multipair territories (chapter seven), though rare and short lived, suggest such a pathway. Excluding the single case of bigamy, we have four good examples of multiple-pair territories. Two of the four cases involved brothers simultaneously pairing with incoming females (case histories 30, 33). In the other two cases, males who had helped more than the usual number of years finally paired and nested within their fathers' territories.

We suspect that the basic dominance system normally prohibits multipair territories in the Florida Scrub Jay. It does appear, however, that the degree of dominance is relatively low among male siblings, and also between a male breeder and a very old male helper. If greater tolerance does exist in these cases, then any slight change in survivorship patterns that would increase the frequency of long-term relationships between brothers, or even fathers and sons, could increase the frequency of multipair territories within a Florida Scrub Jay population. Based on our best example (case history 33), helpers feeding at both nests and

346

breeders feeding fledglings from both nests can arise easily once two successful nests exist simultaneously. The major behavioral shift we have not yet observed involves breeders visiting each other's nests. Once again, this probably stems ultimately from a relaxation of dominance relationships between associating members of the group. (As a contrary point of view, Zahavi [personal communication] argues that feeding at the nest represents an *increased* expression of status.)

The demographic shift that we think could lead to more frequent multipair territories need not involve increased longevity within the population at large. It might, for example, come about through increased fledgling survivorship among only the larger families, thereby conferring an extra advantage to the bigger groups, as well as producing more multipair associations within them. This development of clans resembles the scenario spelled out by Brown (1974), although it is independent of the close inbreeding suggested in that paper. We suspect that detailed information on the demography of Mexican Jays, especially at localities where group sizes are smaller than in southern Arizona, might provide the data needed to test the picture we have developed here.

10.5. RECENT AND FUTURE DIRECTIONS

The list of bird species reported to exhibit cooperative breeding continues to grow rapidly. At present it probably includes several hundred species. Unfortunately, still with rare exceptions, little or nothing is known about these birds beyond someone having observed helpers at the nest. We agree fully with Reyer (1980) that establishing correlations between social structure and environment, while not itself a sufficient end, is a necessary prerequisite for understanding the evolutionary development of social behavior. Such correlations can be generated only with detailed, long-term demographic data.

Despite the diversity of cooperative-breeding systems, we continue to hypothesize that a few underlying ecological similarities unite most or all of them. Our working hypothesis is that competition for, and defense of, some strictly limited resource provides this unification. Indeed, the preliminary information available for many of the better studied bird species often suggests strong similarities to the system as we interpret it among *Aphelocoma* jays. Recent examples include the numerous cooperatively breeding species of New World jays studied by Hardy, Raitt, and their students (e.g., Hardy et al., 1981; Winterstein, 1980), the Galapagos Mockingbird (Kinnaird and Grant, 1982), the Green Woodhoopoe (Ligon and Ligon, 1982), and the Splendid Wren (Rowley, 1981). In the most recent organization of the major characteristics of the best-studied cooperative-breeding birds, Brown (1978a) points out that the majority show many of the characteristics of the Florida Scrub Jay system.

Certain cooperative-breeding systems continue to bewilder us, perhaps because long-term information of the fates of individuals is lacking (e.g., Noisy Miner: Dow, 1978, 1979; Chimney Swift: Dexter, 1952; Barn Swallow: Myers and Waller, 1977; Pinyon Jay: Balda and Balda, 1978; certain species of African starlings: Huels, 1981).

Even in a well-studied species, new information suddenly can provide dramatically new insight. A fine example is the recently published information on the White-fronted Bee-eater (Hegner et al., 1982). For some time the species was known to breed cooperatively in colonies, with many pairs having nonbreeding helpers (Fry, 1972b; Emlen, 1978). Only after years of close study was it discovered that the

348

bee-eater clans, which are extended family groups consisting of a few pairs and their helpers, defend foraging territories (Hegner et al., 1982)! It still is claimed that nest sites are not limited, though this has yet to be demonstrated. Regardless, foraging sites, or foraging sites an optimal distance or direction from the nest, indeed may be limiting. If so, then the ecological saturation model proposed for the Florida Scrub Jay could pertain to this species' once bewildering social system. The fact that territories exist at all certainly suggests that something is in limited supply. If a limiting factor exists, then helpers may be those individuals who, because of youth, widow status, infirmity, or even bad luck, cannot control a sufficient supply of the resources to breed. By remaining in their clan they may increase their fitness directly by enhancing their own opportunities to breed, and indirectly by increasing the representation of close kin other than descendants in future populations. It remains unclear even in this well-studied case how the relative magnitudes of direct versus indirect benefits compare.

Clearly, we continue to focus on defense of a limited resource as fundamental to the evolution of cooperative breeding in birds. If gaining access to these resources (be they land, food, mates, or combinations of these and other factors) requires time well beyond the stage of nutritional independence, we propose that helping represents one of the most important modes of direct access among birds. Therefore, we stress the importance of defining and measuring the limits to these *resources* among populations of cooperative breeders. Only with such measures can the resource limitation model be tested. This remains the single most important problem confronting all field studies of cooperative breeders.

Our own field work presently is focused on more precise

analyses of the territories that Florida Scrub Jays use and defend. The relationships between habitat quality, prey density, and territory size must be studied longitudinally as well as comparatively at any one time. We include the effects of seasonality in these studies, especially regarding the mysteries of the acorn harvest and the flush of insects coinciding with the breeding season. With a better grasp on the jays' nutritional needs and the levels of available resources they perceive, we expect to test our notions about the impacts of the social environment upon territory size. Perhaps, someday, our data on habitat quality can be compared to similar data from habitats that are used *and unused* by Scrub Jays in western North America, where cooperative breeding is absent.

We do not expect that all cooperative-breeding birds necessarily can be understood using our framework, which basically is the same framework as that proposed by Koenig and Pitelka (1981) and Emlen (1982a). However, all of us await the day that sufficient field data emerge, from a large enough variety of studies, even to be able to ask the question.

Epilogue

A decade of field work provides the base data for this book. Financial support and time off from other duties afforded us the opportunity to begin analyzing an extensive data base and preparing a manuscript in spring, 1979. One of our first decisions was to analyze data through the 1979 breeding season but not beyond. (At this writing, July 1982, we have data on three more years of nesting, 1980–1982, and associated survival.) Our decision proved to coincide with a biological, as well as a logistical, landmark.

During the 10 years, monthly or more frequent censusing of the jays permitted us to determine that breeder survival varied from 69 to 93 percent per annum $\bar{x} = 82$), and fledgling survival over the first twelve months varied from 21 to 44 percent ($\bar{x} = 34$). In 1979, after a highly successful nesting season, an unprecedented increase in mortality swept through the population, affecting especially the breeders and juveniles. Survival of breeders between the 1979 and 1980 breeding seasons plummeted to 55 percent (only 39 of 71 breeders survived). Fledgling survival through the same period was near zero: a record 93 fledglings were produced in 1979, of which only 1 was alive by May 1980. Mortality rates returned to normal after March of 1980.

Although we found neither carcasses nor sick jays, we are convinced that of the jays who disappeared in such overwhelming numbers between August 1979 and February 1980, essentially all died. In late fall, after a two-month peak in the disappearance rate, we searched about

16 square kilometers of surrounding scrub for possible dispersers. We found none of our missing jays.

The death of roughly 130 jays (80 more than would be expected normally) was not a result of starvation. As usual, the autumn acorn crop was plentiful. We captured a sample of adults and without exception found their weights to be normal. Furthermore, our tame jays know humans as a peanut source. Any tame jays that were starving would have swarmed over us to seek nourishment, but this did not happen. Some natural element besides starvation killed the jays. Predation and disease are the only remaining possibilities. We suspect the latter, although we never were able to positively confirm the existence of any epizootic agent.

No increase in the activity of any potential jay predator was reported at any time during the critical fall period, and an experienced observer was afield with the jays almost daily at this time. One environmental factor *was* highly peculiar during fall, 1979. Water levels in the study tract reached previously unattained heights following unusually heavy late summer rains. At the peak levels, in late September, the grassy depressions were brim full and even flooding adjacent low areas; whitecaps were seen on the usually dry grassy depression we call Big Lake. Still, the jays' favored oak habitat remained dry, and no crowding of their population would have occurred.

Our best guess is that the unusual water conditions, combined with generally high population levels (see figure 3.18), provided favorable conditions for the nurture and transmission of some disease. We know from individual cases at other times that sick jays become very secretive; hence we would not expect to see unhealthy individuals. Carcasses would last only a few days at most, before being dismembered and devoured by a host of scavengers.

The changes these stupendous losses caused in the population were significant, but remarkably short lived. From August 1979 through February 1980, each monthly census revealed new absences among previously vigorous breeders and nonbreeders. The absent jays were never seen again. Re-pairing and territorial boundary adjustments would begin, only to be struck down by an ensuing loss. However, despite a 45 percent loss of breeders, their density in March 1980 had dropped only 27 percent (from 7.40 to 5.40 breeders per 40 ha), as older helpers moved in to fill the ranks. Although the spate of deaths terminated by February, the breeding season was delayed two or three weeks beyond its normal early March onset.

Many territorial boundaries remained in hot dispute longer than usual. Most pairs attempted only one nesting, perhaps because most pair bonds were only newly formed, still-fragile unions between survivors of "the epidemic."

By June, the late-forming pairs had solidified their bonds enough to have nested and overall breeder density increased to 6.64 (breeders per 40 ha), a level frequently reached during the preceding decade (figure 3.18). Only a few older helpers remained in their natal territories, and just one yearling helper was alive within the tract. (Typical juvenile survival would have left 33!) Reproductive success was lower than usual in 1980, but not dramatically so.

By 1981, the only tangible evidence, besides in the record books, for the precipitous loss of two-thirds of our marked jays was a lower-than-usual population of older helpers. An entire cohort, save one individual and a few immigrants, was missing from the age structure of our population. The "epidemic" and its aftermath left us convinced of two thoughts with which we conclude this book:

1. The remarkable speed with which a severely decimated breeding population returned to its previous, stable level supports our suspicion (chapter three) that much of

the biology of the Florida Scrub Jay population revolves around the attainment of breeding status within a crowded environment. A population of nonbreeding jays waits in the wings at all times, with individuals doing what they can to join the breeding ranks. We see the helper system that characterizes these jays, not as a separate strategy for genetically indirect reproduction, but, instead, intimately involved with the continual process of survival and dispersal. That competition for breeding space is severe and ever present can be confirmed in no better way than to observe a major fraction of the surviving nonbreeders refill the breeding ranks essentially as fast as the breeders disappear, even during the heart of a population crash.

2. *Long-term* is a relative term when used to describe population studies. In just a few years we were able to establish many of the gross features of Florida Scrub Jay biology. After 10 years, we felt we knew enough to present a summary of such population parameters as age structures, life tables, and long-term variance in all-important survivorship and reproductive parameters. The population crash of 1979 left us considerably less cocksure about how much we really do understand the "long-term" biology of our population. With what frequency do these events happen? Are they cyclic or random through time? Do they always affect the oldest and youngest age classes most strongly? Such questions have profound impact on the selective regimes determining life history optima, and we still are largely in the dark about them. We hope someday not to be, because this study will continue indefinitely.

The more we know about Florida Scrub Jays, the more we understand how much we still need to measure, especially regarding annual variations in reproduction, survival, and dispersal parameters. It has been impressed upon us that many of the graphs we need to produce and analyze contain only one data point per year of study. In this vein, it sobers us to see how much of modern avian ecology is based upon studies of only a few years' duration. In relation to such studies, our approach has indeed been of "long term." However, we wonder how much more we will know about the social systems and evolution of birds when a number of studies, ours included, finally do achieve a level of duration that actually *is* long with respect to the generation times of the birds involved. To that end, this book represents just a beginning.

Appendixes

Gross Habitat Features of the Study Tract

For orientation at the Archbold Biological Station, we show the main laboratory building (black rectangle), the west boundary fence, and the railroad that runs roughly north-south through the Station property. Habitat key: (1) prime oak scrub, mostly along the higher sand ridges; (2) transitional, mixed oak-palmetto-wiregrass vegetation; (3) grassy, interdunal depressions, seasonally flooded, typically bordered by dense rim of saw palmetto and gallberry thickets; (4) dense, mature sand pine scrub with closed or nearly closed canopy; (5) low, slash pine flatwoods with grass and palmetto understory, nearly closed canopy; (6) lush bayhead, 8–12 m tall, with thick, humid understory and entirely closed canopy. Florida Scrub Jays favor habitat (1), forage sporadically in habitat (2), rarely in habitats (3) and (4), essentially never enter habitats (5) and (6). Habitat numbers (1) and (2) correspond to the same numbers in text and figure 3.4; others do not. See figure 3.3 for an aerial photograph of the same area mapped here. This map traced from detailed vegetation map of Abrahamson et al. (in press).

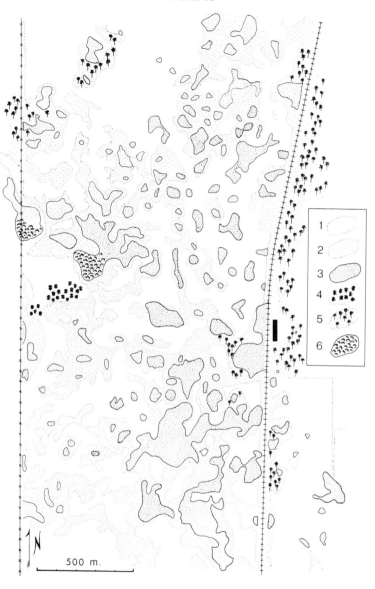

Florida Scrub Jay Territories, 1970–1979

For orientation at the Archbold Biological Station we show the main laboratory building, the west boundary fence, and the railroad. Also shown are major grassy depressions (stipple), which are little used and often form segments of territory boundaries, and two bayhead forests (trees), which never are used. The territory boundaries are marked with solid lines, except where uncertain (dashed). Segments of the same territory, isolated by unused habitat, are joined by arrows. In 1975 and 1976 two patches of the study tract, usually occupied but poor in quality, were not included in any territory (hatched areas). All nesting attempts during each year are mapped accurately. Seasonal first (solid dots), second (open circles), and third (dotted circles) nests are distinguished. A unique fourth attempt (1972, POWL) is so indicated. Occasionally no nest occurred in a territory during a particular year (e.g., 1972, OBEN). The several territories of the same year with more than one nest of the same seasonal attempt were occupied by two pairs (e.g., 1975, 1976, ECUL), except for the one case of bigamy (1974, YTRE). Once (1976, XRDS) a helper son paired with a disperser, and they built a nest in his father's territory (open triangle). During incubation they deserted, and she went home (SIXS). The four-letter symbols are unique to each territory and are used as references to these maps in the text.

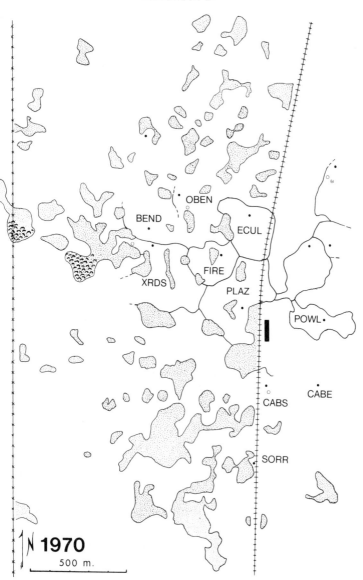

OBEN

BEND

ECUL

FIRE

XRDS

PLAZ

POWL

CABS

CABE

SORR

N 1970

500 m.

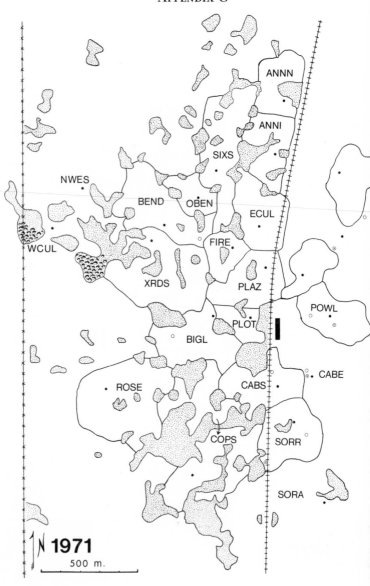

ANNN

ANNI

SIXS

NWES

BEND OBEN

ECUL

WCUL

FIRE

XRDS

PLAZ

POWL

PLOT

BIGL

ROSE

CABS

CABE

COPS

SORR

SORA

N 1971

500 m.

362

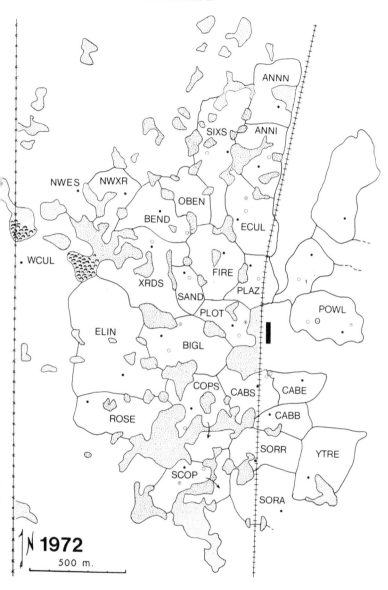

ANNN

SIXS ANNI

NWES NWXR

OBEN

BEND ECUL

WCUL

XRDS FIRE

SAND PLAZ

PLOT POWL

ELIN

BIGL

COPS CABS CABE

ROSE CABB

SORR YTRE

SCOP

SORA

N 1972

500 m.

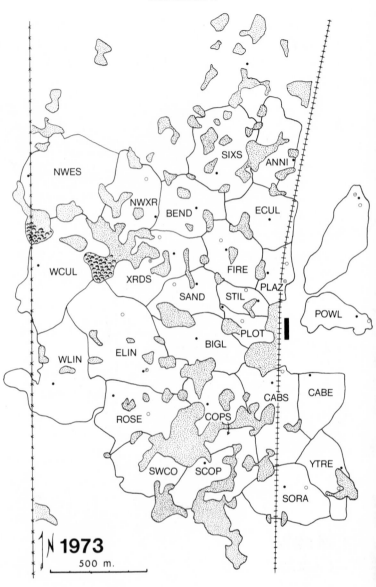

N 1973
500 m.

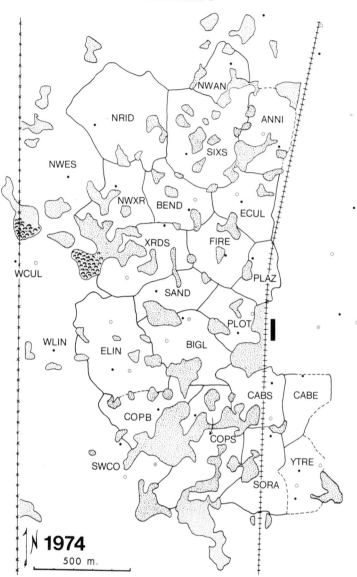

1974

500 m.

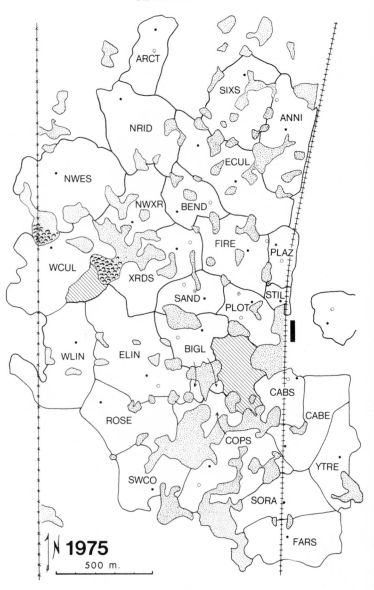

1975

500 m.

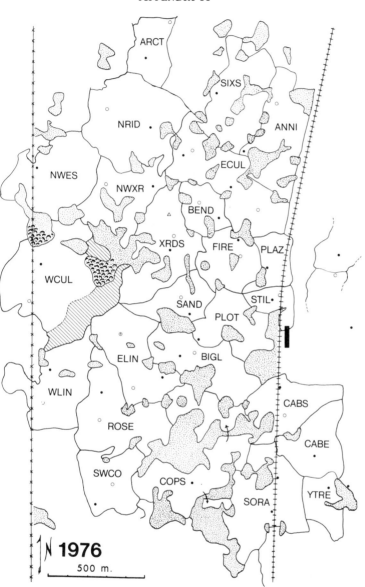

N **1976**

500 m.

367

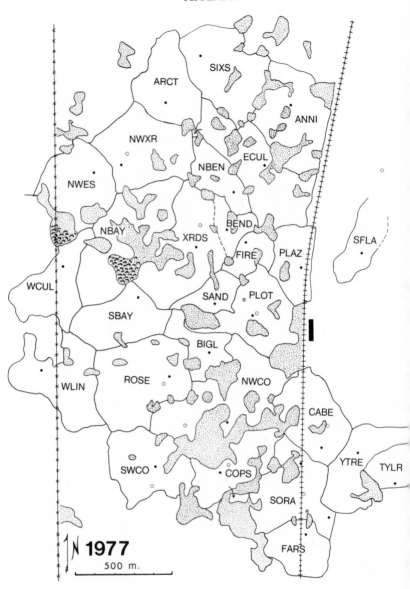

1977

500 m.

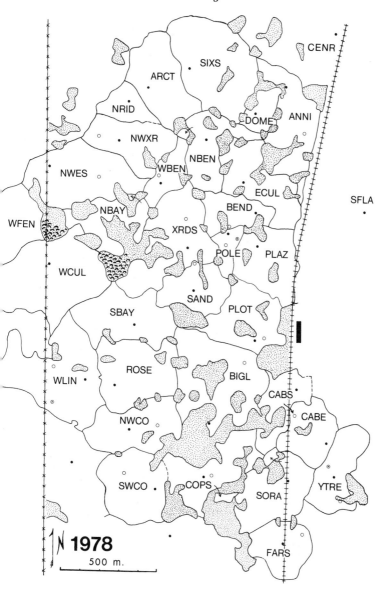

CENR

SIXS

ARCT

NRID

DOME

ANNI

NWXR

NBEN

NWES

WBEN

ECUL

SFLA

NBAY

BEND

WFEN

XRDS

POLE

PLAZ

WCUL

SAND

SBAY

PLOT

ROSE

BIGL

WLIN

CABS

NWCO

CABE

SWCO

COPS

SORA

YTRE

1978

FARS

500 m.

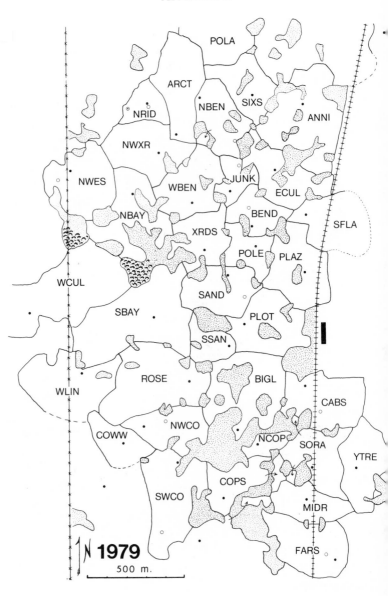

1979

500 m.

Close Inbreeding Among Successful Jays

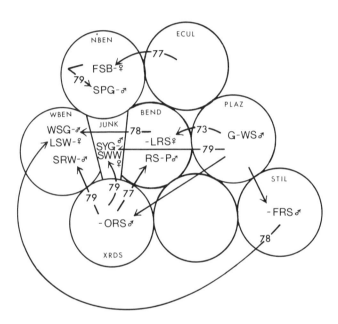

The geography and chronology of close inbreeding. Arrows point *to* the jays that dispersed, and originate *in* the territories from which these jays dispersed. Year of dispersal is given for the nine jays involved in close inbreeding. Any two jays of opposite sex shown in the same territory were paired and produced fertile eggs. See chapter four (4.6 Close Inbreeding), table L.1 (next page), and territory maps (Appendixes E-K) for additional details.

TABLE L.1. Known Breeders Descended from the Most Successful Breeder (G–WS♂) during the Years 1969–1979. (See also figure on preceding page.)

Territory	Years		Individual Scrub Jays		
PLAZ	196?–1976	Pair:	G–WS♂	×	(Y–YS♀)
	–1969	Offspring:	LL–S♂		AA–S♀
	–1970		–ORS♂		–FRS♀, –LRS♂
	–1971		BWS–♀		
	–1976		–BSR♂		
PLAZ	1977 only	Pair:	above	×	(AS–W♀)
		Offspring:	SYG–♂		
XRDS	1972–1979	Pair:	–ORS♂	×	(B–PS♀)
	–1972	Offspring:	RS–P♂		
	–1975		WSW–♂		RSW–♀
	–1976		–YSW♀		
	–1977		SRW–♂		SWW–♀
PLOT	1974–1980	Pair:	LL–S♂	×	(BGS–♀)
		Offspring:	SA–G♂		
STIL	1975–1976	Pair:	–FRS♂	×	(AS–W♀)
		Offspring:	LSW–♀		
BEND	1973–1975	Pair:	(B–RS♂)	×	–LRS♀
	–1975	Offspring:	WSG–♂		
BEND	1977 only	Pair:	RS–P♂	×	above
WCUL	1975–1979	Pair:	(W–AS♂)	×	AA–S♀
		Offspring:	WSB–♀		
NBEN	1977–1978	Pair:	WSW–♂	×	(FSB–♀)
		Offspring:	SPG–♂		
NBEN	1979 only	Pair:	SPG–♂	×	above
WBEN	1978 only	Pair:	WSG–♂	×	LSW–♀
WBEN	1979 only	Pair:	SRW–♂	×	above
JUNK	1979 only	Pair:	SYG–♂	×	SWW–♀

NOTE: Descendants are listed as members of a pair *only* if they produced a *breeder* or if they paired with another descendant of G–WS♂. Jays not related to G–WS♂ are shown in parentheses. The word "above" refers to breeder of like sex of preceding pair. As of June 1982, 5 of the 27 individuals still are alive (W–AS♂, SYG–♂, SPG–♂, BGS–♀, SWW–♀).

Calculation of
Helper Death Rates

In an average year, the number of helpers greatly exceeds the number of breeding vacancies created by deaths among mated pairs. Therefore, only a fraction of the helpers can become breeders each year. The remainder either continue as helpers, or they die. For each age and sex class, i, of helpers, then, three fates are possible at age class $i+1$: breeding, helping, or dead.

Ideally, the number of helpers (D) neither breeding nor helping in the study tract at age $i + 1$ would be exactly the number dead. Dividing by the original helper sample (H_i) would yield the helper death rate (m_i) for that age and sex class:

$$m_i = \frac{D}{H_i}$$

However, a small number of helpers can be expected to have dispersed to breed outside our study tract. These must be subtracted from D in the equation above. We arrive at this correction factor by knowing what proportion of i age helpers should fill the known number of vacancies left by an annual breeder mortality (d) among a known sample (B) of breeders:

$$m_i = \frac{D - (a{\cdot}B{\cdot}d)}{H_i}$$

where a = known proportion of new breeders from age class i. However, the quantity ($a \cdot B \cdot d$) will include *all* the breeders age class i, not just those outside the study tract. We must subtract from this quantity the number, b, of i age helpers *already known* to be breeders and therefore never included within any estimate or actual count of D:

$$m_i = \frac{D - (a \cdot B \cdot d - b)}{H_i}$$

Tables M.1 and M.2 present the pertinent data for our calculations. The figures are from 1970 through 1978, and they exclude 7 yearling helpers whose sex was never established before they disappeared. Table 9.5 presents the maximum possible range of values for yearling helper death rates, assuming these 7 to be either all males or all females, *and* assuming no successful dispersal outside our tract. Here, we correct these figures downward using the calculation derived above and the values in tables M.1 and M.2, which follow.

Yearling Males:

$$m_i = \frac{9 - (.47 \times 252 \times .18 - 23)}{64}$$
$$= \frac{9 - (\text{``0''})}{64}$$

minimum corrected m_i = .141

maximum corrected m_i (assuming 7 unknowns were males):

$$\frac{16}{71} = .225$$

Maximum possible range (see table 9.5): .141–.225
Corrected range (calculations above): .141–.225
Probable value (most unknowns = males): about .20–.22

Yearling Females:

$$m_i = \frac{34 - (.63 \times 252 \times .18 - 21)}{81}$$

minimum corrected m_i = .326

maximum corrected m_i (assuming 7 unknowns were females):

$$m_i = \frac{41 - (.63 \times 252 \times .18 - 21)}{88}$$
$$= .380$$

Maximum possible range (see table 9.5): .420–.466
Corrected range (calculations above): .326–.380
Probable value (most unknowns = males): about .35

Two-Year-Old Males:

$$m_i = \frac{7 - (.28 \times 235 \times .18 - 9)}{28}$$
$$= .148$$

Maximum possible: $\frac{7}{28} = .25$

Corrected value (above): .15

Two-Year-Old Females:

$$m_i = \frac{9 - (.35 \times 235 \times .18 - 14)}{24}$$
$$= .341$$

Maximum possible: $\frac{9}{24} = .375$

Corrected value (above): .34

Appendix M

Table M.1. Fates of Female Helpers One Year after Helping, 1970–1978

| | | Yearling Females | | | | Two-Year-Old Females | | | |
| | | | Subsequent year | | | | Subsequent year | | |
Year	Female Breeders (B)	Helpers (H_i)	Breeding (b)	Helping (h)	D[d]	Helpers (H_i)	Breeding (b)	Helping (h)	D
1970	17	5	3	2	0	—	—	—	—
1971	27	10	3	4	3	2	2	0	0
1972	33	5	2	1	2	4	3	0	1
1973	28	3	1	1	1	1	0	0	1
1974	28	11	2	2	7	1	1	0	0
1975	26	4	1	3	0	2	1	1	0
1976	29	20	4	7	11	3	2	0	
1977	30	9	0	4	6	7	2	0	5
1978	34	14	5	2	7	4	3	0	
Total	252[a]	81	21	26	34	24	14	1	9
Ratio to breeding population[b]		.321	—	—	—	.095	—	—	—
Proportional entry into breeding pop.[c]		.63				.35			

[a] No known-age two year olds were available in 1970; total sample of breeding adult becomes 235 (252 − 17) for two-year-old helper calculations.
[b] Ratio to breeding population = number of helpers / number of breeders.
[c] Proportional entry into breeding population by helpers of different ages, taken from table 4.1.
[d] D = Disappeared from tract through Dispersal or Death.

Table M.2. Fates of Male Helpers One Year after Helping, 1970–1978

Year	Male Breeders	Yearling Males				Two-Year-Old Males			
		Helpers	Subsequent year			Helpers	Subsequent year		
			Breeding	Helping	D[d]		Breeding	Helping	D[d]
1970	17	3	1	2	0	—	—	—	—
1971	27	12	5	5	2	2	0	1	1
1972	33	7	0	6	1	5	1	3	1
1973	28	5	0	3	2	6	0	5	1
1974	28	2	1	0	1	3	1	2	0
1975	26	11	4	5	2	0	0	0	0
1976	29	10	4	6	0	5	4	0	1
1977	30	4	2	1	1	6	3	1	2
1978	34	10	6	4	0	1	0	0	1
Total	252[a]	64	23	32	9	28	9	12	7
Ratio to breeding population[b]	.254	—	—	—	.111	—	—	—	
Proportional entry into breeding pop.[c]	.47					.28			

NOTE: Footnotes as in table M.1.

Scientific Names
of Cooperative Breeders

TABLE N.1. Scientific Names of Cooperative Breeding Birds Mentioned in Text

Family	Common Name	Scientific Name
Accipitridae	Harris Hawk	*Parabuteo unicinctus*
	Galapagos Hawk	*Buteo galapagoensis*
Rallidae	Tasmanian Native Hen	*Tribonyx morterii*
	Pukeko	*Porphyrio porphyrio*
Cuculidae	Groove-billed Ani	*Crotophaga sulcirostris*
Apodidae	Chimney Swift	*Chaetura pelagica*
Alcedinidae	Pied Kingfisher	*Ceryle rudis*
	Kookaburra	*Dacelo novaeguineae*
Todidae	Puerto Rican Tody	*Todus mexicanus*
Meropidae	Red-throated Bee-eater	*Merops bullocki*
	White-fronted Bee-eater	*Merops bullockoides*
Phoeniculidae	Green Woodhoopoe	*Phoeniculus purpureus*
Picidae	Acorn Woodpecker	*Melanerpes formicivorus*
	Red-cockaded Woodpecker	*Picoides borealis*
Tyrannidae	White-bearded Flycatcher	*Myiozetetes inornatus*
Hirundinidae	Barn Swallow	*Hirundo rustica*
Mimidae	Galapagos Mockingbird	*Nesomimus trifasciatus*
Prunellidae	Dunnock	*Prunella modularis*
Muscicapidae	Grey-crowned Babbler	*Pomatostomus temporalis*
	Hall's Babbler	*Pomatostomus halli*
	Common Babbler	*Turdoides caudatus*
	Arabian Babbler	*Turdoides squamiceps*
	Jungle Babbler	*Turdoides striatus*
	Superb Blue Wren	*Malurus cyaneus*
	Splendid Wren	*Malurus splendens*
Aegithalidae	Long-tailed Tit	*Aegithalos caudatus*
Meliphagidae	Noisy Miner	*Manorhina melanocephala*
Cracticidae	Black-backed Magpie	*Gymnorhina tibicen*
Corvidae	Pinyon Jay	*Gymnorhinus cyanocephalus*
	Scrub Jay	*Aphelocoma coerulescens*
	Mexican Jay	*Aphelocoma ultramarina*
	Unicolored Jay	*Aphelocoma unicolor*
	Brown Jay	*Psilorhinus morio*

References

Abrahamson, W. G., A. F. Johnson, J. N. Layne, and P. A. Peroni. Vegetation of the Archbold Biological Station, Florida: An Example of the Southern Lake Wales Ridge. *Florida Scientist,* in press.

Alexander, R. D. 1974. The evolution of social behavior. *Ann. Rev. Ecol. & Syst.* 5: 325–383.

Allen, J. A. 1871. On the mammals and winter birds of east Florida, with an examination of certain assumed specific characters in birds, and a sketch of the bird-faunae of eastern North America. *Bull. Mus. Comp. Zool.* 2: 161–450.

Alvarez, H. 1975. The social system of the Green Jay in Colombia. *Living Bird* 14: 5–44.

Amadon, D. 1944. A preliminary life history study of the Florida Jay, *Cyanocitta c. coerulescens. Amer. Mus. Novitates* no. 1252, 22 pp.

Ashmole, N. P. 1971. Sea bird ecology and the marine environment. In: Farner, D. S., and J. R. King, eds. *Avian Biology,* vol. 1, pp. 224–286. New York: Academic Press.

Atwood, J. L. 1978. The breeding biology of the Santa Cruz Island Scrub Jay, *Aphelocoma coerulescens insularis.* M.A. thesis, California State Univ., Long Beach.

Atwood, J. L. 1980a. Social interactions in the Santa Cruz Island Scrub Jay. *Condor* 82: 440–448.

Atwood, J. L. 1980b. Breeding biology of the Santa Cruz Island Scrub Jay. In: Power, D. M., ed. *The California Islands: Proceedings of a Multidisciplinary Symposium,* pp. 675–688. Santa Barbara: Santa Barbara Mus. Nat. Hist.

REFERENCES

Axelrod, D. I. 1958. Evolution of the Madro-Tertiary geo-flora. *Bot. Rev.* 24: 433–509.

Baber, D. W., and J. G. Morris. 1980. Florida Scrub Jays foraging from feral hogs. *Auk* 97: 202.

Balda, R. P., and J. H. Balda. 1978. The care of young Piñon Jays (*Gymnorhinus cyanocephalus*) and their integration into the flock. *J. f. Ornithol.* 119: 146–171.

Balph, D. F., G. S. Innis, and M. H. Balph. 1980. Kin selection in Rio Grande Turkeys: a critical assessment. *Auk* 97: 854–860.

Bancroft, G. T., and G. E. Woolfenden. 1982. The molt of Scrub jays and Blue Jays in Florida. *Ornithol. Monogr.* no. 29, 51 pp.

Barbour, D. B. 1977. Vocal communication in the Florida Scrub Jay. M.A. thesis, Univ. of South Florida.

Barrowclough, G. F. 1978. Sampling bias in dispersal studies based on finite areas. *Bird-Banding* 49: 333–341.

Barrowclough, G. F. 1980. Gene flow, effective population sizes, and genetic variance components in birds. *Evol.* 34: 789–798.

Bendire, C. E. 1895. *Life Histories of North American Birds, from Parrots to Grackles.* Washington, D.C.: Gov't. Printing Office.

Bent, A. C. 1946. Life histories of North American jays, crows, and titmice. *U.S. Nat. Mus. Bull.*, no. 191.

Bertram, B. C. 1978. Living in groups: predators and prey. In: Krebs, J. R., and N. B. Davies, eds. *Behavioural Ecology: An Evolutionary Approach*, pp. 64–96. Oxford: Blackwell.

Birkhead, M. E. 1981. The social behaviour of the Dunnock, *Prunella modularis. Ibis* 123: 75–84.

Botkin, D. B., and R. S. Miller. 1974. Mortality rates and survival of birds. *Amer. Nat.* 108: 181–192.

Brockman, C. F. 1968. *Trees of North America.* New York: Golden Press.

Brown, J. L. 1963. Social organization and behavior of the Mexican Jay. *Condor* 65: 126–153.

Brown, J. L. 1964. The evolution of diversity in avian territorial systems. *Wilson Bull.* 76: 160–169.

Brown, J. L. 1969. Territorial behavior and population regulation in birds. *Wilson Bull.* 81: 293–329.

Brown, J. L. 1970. Cooperative breeding and altruistic behavior in the Mexican Jay, *Aphelocoma ultramarina. Anim. Behav.* 18: 366–378.

Brown, J. L. 1972. Communal feeding of nestlings in the Mexican Jay (*Aphelocoma ultramarina*): interflock comparisons. *Anim. Behav.* 20: 395–403.

Brown, J. L. 1974. Alternate routes to sociality in jays— with a theory for the evolution of altruism and communal breeding. *Amer. Zool.* 14: 63–80.

Brown, J. L. 1975a. *The Evolution of Behavior.* New York: Norton.

Brown, J. L. 1975b. Helpers among Arabian Babblers, *Turdoides squamiceps. Ibis* 117: 243–244.

Brown, J. L. 1978a. Avian communal breeding systems. *Ann. Rev. Ecol. & Syst.* 9: 123–156.

Brown, J. L. 1978b. Avian heirs of territory. *BioScience* 28: 750–752.

Brown, J. L. 1980. Fitness in complex avian social systems. In: Markl, H., ed. *Evolution of Social Behavior: Hypotheses and Empirical Tests*, pp. 115–128. Deerfield Beach, Fla.: Verlag Chemie.

Brown, J. L., and R. P. Balda. 1977. The relationship of habitat quality to group size in Hall's Babbler, *Pomatostomus halli. Condor* 79: 312–320.

Brown, J. L., and E. R. Brown. 1981a. Extended family system in a communal bird. *Science* 211: 959–960.

Brown, J. L., and E. R. Brown. 1981b. Kin selection and individual selection in babblers. In: Alexander, R. D., and D. Tinkle, eds. *Natural Selection and Social Behavior: Recent Research and New Theory*, pp. 244–256. New York: Chiron Press.

Brown, J. L., E. R. Brown, S. D. Brown, and D. D. Dow. 1982. Helpers: Effects of experimental removal on reproductive success. *Science* 215: 421–422.

Brown, J. L., and G. H. Orians. 1970. Spacing patterns in mobile animals. *Ann. Rev. Ecol. & Syst.* 1: 239–262.

Buitron, D. 1983. Extra-pair courtship in Black-billed Magpies. *Anim. Behav.* 31: 211–220.

Bulmer, M. G., and C. M. Perrins. 1973. Mortality in the Great Tit *Parus major. Ibis* 115: 277–281.

Caccamise, D. F. 1976. Nesting mortality in the Red-winged Blackbird. *Auk* 93: 517–534.

Carrick, R. 1972. Population ecology of the Australian Black-backed Magpie, Royal Penguin, and Silver Gull. In: *Population Ecology of Migratory Birds: A Symposium*, pp. 41–99. U.S. Dept. of Interior Wildl. Res. Rept. 2.

Cody, M. L. 1966. A general theory of clutch size. *Evol.* 20: 174–184.

Cohen, D. 1966. Optimizing reproduction in a randomly varying environment. *J. theoret. Biol.* 12:119–129

Cole, L. C. 1954. The population consequences of life history phenomena. *Quart. Rev. Biol.* 29: 103–137

Coulson, J. C. 1966. The influence of the pair-bond and age on the breeding biology of the Kittiwake gull *Rissa tridactyla. J. Anim. Ecol.* 35: 269–279.

Coulson, J. C., and R. D. Wooler. 1976. Differential survival rates among breeding Kittiwake gulls *Rissa tridactyla* (L.). *J. Anim. Ecol.* 45: 205–213.

Cox, J. A. 1983. Bartram's bird, the Florida Scrub Jay. *Florida Nat.* 56: 7–10.

REFERENCES

Craig, J. L. 1976. An interterritorial hierarchy: an advantage for a subordinate in a communal territory. *Zeitschrift f. Tierpsych.* 42: 200–205.

Craig, J. L. 1977. The behaviour of the Pukeko, *Porphyrio porphyrio melanotus. New Zealand J. Zool.* 4: 413–433.

Craig, J. L. 1979. Habitat variation in the social organization of a communal gallinule, the Pukeko, *Porphyrio porphyrio melanotus. Behav. Ecol. & Sociobiol.* 5: 331–358.

Craig, J. L. 1980a. Pair and group breeding behavior of a communal gallinule, the Pukeko, *Porphyrio p. melanotus. Anim. Behav.* 28: 593–603.

Craig, J. L. 1980b. Breeding success of a communal gallinule. *Behav. Ecol. & Sociobiol.* 6: 289–295.

Crossin, R. S. 1967. The breeding biology of the Tufted Jay. *Proc. Western Found. Vertebr. Zool.* 1: 265–299.

Darwin, C. R. 1859. *On the Origin of Species.* London: Murray.

Davies, N. B. 1978. Ecological questions about territorial behavior. In: Krebs, J. R., and N. B. Davies, eds. *Behavioural Ecology: An Evolutionary Approach*, pp. 317–350. Oxford: Blackwell.

Davies, N. B. 1982. Cooperation and conflict in breeding groups. *Nature* 296: 702–703.

Deevey, E. S., Jr. 1947. Life tables for natural populations of animals. *Quart. Rev. Biol.* 22: 283–314.

DeGange, A. R. 1976. The daily and annual time budget of the Florida Scrub Jay. M.A. thesis, Univ. of South Florida.

DeSteven, D. 1978. The influence of age on the breeding biology of the Tree Swallow *Iridoprocne bicolor. Ibis* 120: 516–523.

Dexter, R. W. 1952. Extra-parental cooperation in the nesting of Chimney Swifts. *Wilson Bull.* 63: 133–139.

Dixon, J. G. 1944. California Scrub Jay picks ticks from mule deer. *Condor* 46: 204.

Dow, D. D. 1978, Reproductive behavior of the Noisy Miner, a communally breeding honeyeater. *Living Bird* 16: 163–185.

Dow, D. D. 1979. Agonistic and spacing behavior of the Noisy Miner *Manorina melanocephala*, a communally breeding honeyeater. *Ibis* 121: 423–436.

Emigh, T. H., and E. Pollak. 1979. Fixation probabilities and effective population numbers in diploid populations with overlapping generations. *Theoret. Pop. Biol.* 15: 86–107.

Emlen, J. M. 1973. *Ecology: An Evolutionary Approach.* Reading, Mass.: Addison–Wesley.

Emlen, S. T. 1978. The evolution of cooperative breeding in birds. In: Krebs, J. R., and N. B. Davies, eds. *Behavioural Ecology: An Evolutionary Approach*, pp. 245–281. Oxford: Blackwell.

Emlen, S. T. 1982a. The evolution of helping behavior. I. An ecological constraints model. *Amer. Nat.* 119: 29–39.

Emlen, S. T. 1982b. The evolution of helping. II. The role of behavioral conflict. *Amer. Nat.* 119: 40–53.

Emlen, S. T., and L. W. Oring. 1977. Ecology, sexual selection and the evolution of mating systems. *Science* 197: 215–223.

Faaborg, J., T. de Vries, C. B. Patterson, and C. R. Griffin. 1980. Preliminary observations on the occurrence and evolution of polyandry in the Galapagos Hawk (*Buteo galapagoensis*). *Auk* 97: 581–590.

Faaborg, J., and C. B. Patterson. 1981. The characteristics and occurrence of cooperative polyandry. *Ibis* 123: 477–484.

Felsenstein, J. 1975. A pain in the torus: some difficulties with models of isolation by distance. *Amer. Nat.* 109: 359–368.

Fisher, R. A. 1930. *The Genetical Theory of Natural Selection.* London: Oxford Univ. Press.

Fitzpatrick, J. W., and G. E. Woolfenden. 1981. Demography is a cornerstone of sociobiology. *Auk* 98: 406–407.

Fry, C. H. 1972a. The social organization of bee-eaters (Meropidae) and cooperative breeding in hot-climate birds. *Ibis* 114: 1–14.

Fry, C. H. 1972b. The biology of African bee-eaters. *Living Bird* 11: 75–112.

Fry, C. H. 1975. Cooperative breeding in bee-eaters and longevity as an attribute of group-breeding birds. *Emu* 75: 308–309.

Fry, C. H. 1977. The evolutionary significance of co-operative breeding in birds. In: Stonehouse, B., and C. Perrins, eds. *Evolutionary Ecology*, pp. 127–135. London: Macmillan & Co.

Gabaldon, D. J., and R. P. Balda. 1980. Effects of age and experience on breeding success in Pinyon Jays (Abs.). *Amer. Zool.* 20: 787.

Gaston, A. J. 1973. The ecology and behaviour of the Long-tailed Tit. *Ibis.* 115: 330–351.

Gaston, A. J. 1977. Social behaviour within groups of Jungle Babblers (*Turdoides striatus*). *Anim. Behav.* 25: 828–848.

Gaston, A. J. 1978a. The evolution of group territorial behavior and cooperative breeding. *Amer. Nat.* 112: 1091–1100.

Gaston, A. J. 1978b. Ecology of the Common Babbler *Turdoides caudatus. Ibis* 120: 415–432.

Gaston, A., and C. Perrins. 1974. The relationship of habitat to group size in the genus *Turdoides*. *16th Internat. Ornithol. Congr. Abst.*, p. 109.

Goodwin, D. 1976. *Crows of the World*. Ithaca, N.Y.: Comstock Publ. Associates.

Greenwood, P. J. 1980. Mating systems, philopatry and dispersal in birds and mammals. *Anim. Behav.* 28: 1140–1162.

Greenwood, P. J., P. H. Harvey, and C. M. Perrins. 1979a. The role of dispersal in the Great Tit (*Parus major*): the causes, consequences and heritability of natal dispersal. *J. Anim. Ecol.* 48: 123–142.

Greenwood, P. J., P. H. Harvey, and C. M. Perrins. 1979b. Mate selection in the Great Tit *Parus major* in relation to age, status and natal dispersal. *Ornis Fennica* 56: 75–86.

Hamilton, W. D. 1964. The genetical evolution of social behavior. I, II. *J. theoret. Biol.* 7: 1–52.

Hamilton, W. D. 1966. The moulding of senescence by natural selection. *J. theoret. Biol.* 12: 12–45.

Hamilton, W. D. 1972. Altruism and related phenomena, mainly in social insects. *Ann. Rev. Ecol. & Syst.* 3: 193–232.

Hardy, J. W. 1961. Studies in behavior and phylogeny of certain New World jays (Garrulinae). *Univ. Kansas Sci. Bull.* 42: 13–149.

Hardy, J. W. 1976. Comparative breeding behavior and ecology of the Bushy-crested and Nelson San Blas jays. *Wilson Bull.* 88: 96–120.

Hardy, J. W., T. A. Webber, and R. J. Raitt. 1981. Communal social biology of the southern San Blas Jay. *Bull. Florida State Mus.* 26: 203–263.

Harvey, P. H., P. J. Greenwood, C. M. Perrins, and A. R. Martin. 1979. Breeding success of Great Tits *Parus major* in relation to age of male and female parent. *Ibis* 121: 216–219.

Hegner, R. E., S. T. Emlen, and N. J. Demong. 1982. Spatial organization of the White-fronted Bee-eater. *Nature* 298: 264–266.

Hickey, J. J. 1952. Survival studies of banded birds. U.S. Dept. of Interior Fish and Wildlife Serv. Spec. Sci. Rept. 15: 1–177.

Hirshfield, M. F., and D. W. Tinkle. 1975. Natural selection and the evolution of reproductive effort. *Proc. Nat. Acad. Sci.* 72: 2227–2231.

Hoogland, J. L. 1982. Prairie dogs avoid extreme inbreeding. *Science* 215: 1639–1641.

Horn, H. S. 1978. Optimal tactics of reproduction and life-history. In: Krebs, J. R., and N. B. Davies, eds. *Behavioural Ecology: An Evolutionary Approach*, pp. 411–429. Oxford: Blackwell.

Howard, H. E. 1920. *Territory in Bird Life*. London: John Murray.

Howell, A. H. 1932. *Florida Bird Life*. New York: Little and Ives Co.

Huels, T. R. 1981. Cooperative breeding in the Golden-breasted Starling *Cosmopsarus regius*. *Ibis* 123: 539–542.

Hutchinson, G. E. 1978. *An Introduction to Population Ecology*. New Haven: Yale Univ. Press.

Jackson, J. F. 1973. Distribution and population phenetics of the Florida scrub lizard, *Sceloporus woodi*. *Copeia* 1973: 746–761.

Johnson, D. H. 1979. Estimating nest success: the Mayfield method and an alternative. *Auk* 96: 651–661.

Joste, N. 1983. An electrophoretic analysis of paternity in the Acorn Woodpecker Abstracts, 53rd Annual Meeting Cooper Ornithol. Soc., Albuquerque, N.M.

Kale, H. W., II, ed. 1978. *Rare and Endangered Biota of*

Florida, vol. 2. Gainesville: University Presses of Florida.

Kepler, A. K. 1977. Comparative study of todies (Todidae) with emphasis on the Puerto Rican Tody, *Todus mexicanus*. *Publ. Nuttall Ornithol. Club* 16: 1–190.

Kinnaird, M. F., and P. R. Grant. 1982. Cooperative breeding in the Galapagos Mockingbird, *Nesomimus parvulus*. *Behav. Ecol. & Sociobiol.* 10: 65–73.

Koenig, W. D. 1981. Reproductive success, group size, and the evolution of cooperative breeding in the Acorn Woodpecker. *Amer. Nat.* 117: 421–443.

Koenig, W. D., R. L. Mumme, and F. A. Pitelka. 1983. Female roles in cooperatively breeding Acorn Woodpeckers. In: Wasser, S. K., ed. *Social Behavior of Female Vertebrates,* pp. 235–261. New York: Academic Press.

Koenig, W. D., and F. A. Pitelka. 1979. Relatedness and inbreeding avoidance: Counterploys in the communally nesting Acorn Woodpecker. *Science* 206: 1103–1105.

Koenig, W. D., and F. A. Pitelka. 1981. Ecological factors and kin selection in the evolution of cooperative breeding in birds. In: Alexander, R. D., and D. Tinkle, eds. *Natural Selection and Social Behavior: Recent Research and New Theory,* pp. 261–280. New York: Chiron Press.

Krebs, J. R. 1970. Regulation of numbers in the Great Tit (Aves: Passeriformes). *J. Zool. London* 162: 317–333.

Lack, D. 1954. *The Natural Regulation of Animal Numbers.* London: Oxford Univ. Press.

Lack, D. 1966. *Population Studies of Birds.* Oxford: Clarendon Press.

Lack, D. 1968. *Ecological Adaptations for Breeding in Birds.* London: Methuen.

Laessle, A. M. 1942. The plant communities of the Welaka area. *Univ. Florida Publ., Biol. Sci. Ser.* IV. 1: 1–143.

Laughlin, R. 1965. Capacity for increase: a useful population statistic. *J. Anim. Ecol.* 34: 77–91.

Lawton, M. F., and C. F. Guindon. 1981. Flocking composition, breeding success, and learning in the Brown Jay. *Condor* 83: 27–33.

Lazarus, J. 1972. Natural selection and the functions of flocking in birds: a reply to Murton. *Ibis* 114: 556–558.

Leslie, P. H. 1945. On the use of matrices in certain population mathematics. *Biometrika* 33: 183–212.

Leslie, P. H. 1948. Some further notes on the use of matrices in population mathematics. *Biometrika* 35: 213–245.

Leslie, P. H. 1966. The intrinsic rate of increase and the overlap of successive generations in a population of guillemots (*Uria aalge* Pont.). *J. Anim. Ecol.* 35: 291–301.

Lewontin, R. C. 1965. Selection for colonizing ability. In: Baker, A. G., and G. L. Stebbins, eds. *The Genetics of Colonizing Species*, pp. 77–94. New York: Academic Press.

Ligon, J. D. 1970. Behavior and breeding biology of the Red-cockaded Woodpecker. *Auk* 87: 255–278.

Ligon, J. D., and S. H. Ligon. 1978. Communal breeding in Green Woodhoopoes as a case for reciprocity. *Nature* 276: 496–498.

Ligon, J. D., and S. H. Ligon. 1979. The communal social system of the Green Woodhoopoe in Kenya. *Living Bird* 17: 159–197.

Ligon, J. D., and S. H. Ligon. 1982. The cooperative breeding behavior of the Green Woodhoopoe. *Sci. Amer.* 247: 126–134.

Lotka, A. J. 1922. The stability of the normal age distribution. *Proc. Nat. Acad. Sci.* 8: 339–345.

MacArthur, R. 1968. Selection for life tables in periodic environments. *Amer. Nat.* 102: 381–383.

MacArthur, R. A., and E. O Wilson. 1967. *The Theory of Island Biogeography.* Princeton: Princeton Univ. Press.

MacRoberts, M. H., and B. R. MacRoberts. 1976. Social organization and behavior of the Acorn Woodpecker in central coastal California. *Ornithol. Monogr.* no. 21, 115 pp.

Mader, W. J. 1979. Breeding behavior of a polyandrous trio of Harris' Hawks in southern Arizona. *Auk* 95: 327–337.

Malcolm, J. R., and K. Marten. 1982. Natural selection and the communal rearing of pups in African wild dogs (*Lycaon pictus*). *Behav. Ecol. & Sociobiol.* 10: 1–13.

Markl, H., ed. 1980. *Evolution of Social Behavior: Hypotheses and Empirical Tests.* Deerfield Beach, Fla.: Verlag Chemie.

May, R. M. 1976. Estimating r: a pedagogical note. *Amer. Nat.* 110: 496–499.

May, R. M. 1977. Optimal life history strategies. *Nature* 267: 394–395.

May, R. M., ed. 1981. *Theoretical Ecology.* 2d ed. Sunderland, Mass.: Sinauer.

Mayfield, H. 1961. Nesting success calculated from exposure. *Wilson Bull.* 73: 255–261.

Maynard Smith, J. 1964. Group selection and kin selection. *Nature* 201: 1145–1147.

Maynard Smith, J. 1971. What use is sex? *J. theoret. Biol.* 30: 319–335.

Maynard Smith, J. 1978. *The Evolution of Sex.* Cambridge: Cambridge Univ. Press.

Maynard Smith, J., and M. G. Ridpath. 1972. Wife sharing in the Tasmanian Native Hen, *Tribonyx mortierii*: a case of kin selection? *Amer. Nat.* 106: 447–452.

Michod, R. E. 1979. Genetical aspects of kin selection: effects of inbreeding. *J. theoret. Biol.* 81: 223–233.

Michod, R. E., and W. W. Anderson. 1980. On calculating

demographic parameters from age frequency data. *Ecol.* 61: 265–269.

Morton, S. R., and G. D. Parry. 1974. The auxiliary social system in Kookaburras: a reappraisal of its adaptive significance. *Emu* 74: 196–198.

Moynihan, M. 1962. The organization and probable evolution of some mixed species flocks of neotropical birds. *Smiths. Misc. Coll.* 143: 1–140.

Mumme, R. L., W. D. Koenig, and F. A. Pitelka. 1983. Mate guarding in the Acorn Woodpecker: within-group reproductive competition in a cooperative breeder. *Anim. Behav.*, 31: 1094–1106.

Myers, G. R., and D. W. Waller. 1977. Helpers at the nest in Barn Swallows. *Auk* 44: 596.

Myers, J. P., P. G. Connors, and F. A. Pitelka. 1979. Territory size in wintering Sanderlings: the effects of prey abundance and intruder density. *Auk* 96: 551–561.

Neill, W. T. 1957. Historical biogeography of present-day Florida. *Bull. Florida State Mus.* 2. 7: 175–220.

Nice, M. M. 1937. Studies in the life history of the Song Sparrow. *Trans. Linn. Soc. N. Y.* 4: 1–247.

Nice, M. M. 1943. Studies in the life history of the Song Sparrow. *Trans. Linn. Soc. N. Y.* 6: 1–328.

Nice, M. M. 1957. Nesting success in altricial birds. *Auk* 74: 305–321.

Nolan, V., Jr. 1978. The ecology and behavior of the Prairie Warbler, *Dendroica discolor*. *Ornithol. Monogr.* no. 26, 595 pp.

Norris, R. A. 1958. Comparative biosystematics and life history of the nuthatches *Sitta pygmaea* and *Sitta pusilla*. *Univ. Calif. Publ. Zool.* 56: 119–300.

Orians, G. H., C. E. Orians, and K. J. Orians. 1977. Helpers at the nest in some Argentine blackbirds. In: Stonehouse, B., and C. Perrins, eds. *Evolutionary Ecology*, pp. 137–151. London: MacMillan & Co.

Parry, V. 1973. The auxiliary social system and its effect on territory and breeding in Kookaburras. *Emu* 73: 81–100.

Pianka, E. R. 1970. On r and K selection. *Amer. Nat.* 104: 592–597.

Pianka, E. R. 1972. r and K selection or b and d selection? *Amer. Nat.* 106: 581–588.

Pianka, E. R., and W. S. Parker. 1975. Age-specific reproductive tactics. *Amer. Nat.* 109: 453–464.

Pitelka, F. A. 1951. Speciation and ecologic distribution in American jays of the genus *Aphelocoma*. *Univ. Calif. Publ. Zool.* 50: 195–464.

Power, H. W. 1980. The foraging behavior of Mountain Bluebirds. *Ornithol. Monogr.* no. 28, 72 pp.

Pulliam, H. R. 1973. On the advantages of flocking. *J. theoret. Biol.* 38: 419–422.

Reyer, H-U. 1980. Flexible helper structure as an ecological adaptation in the Pied Kingfisher (*Ceryle rudis*). *Behav. Ecol. & Sociobiol.* 6: 219–229.

Ricklefs, R. E. 1968. On the limitation of brood size in passerine birds by the ability of adults to nourish their young. *Proc. Nat. Acad. Sci.* 61: 847–851.

Ricklefs, R. E. 1969. An analysis of nesting mortality in birds. *Smiths. Contrib. Zool.* 9: 1–48.

Ricklefs, R. E. 1973. Fecundity, mortality, and avian demography. In: Farner, D. S., ed. *Breeding Biology of Birds*, pp. 366–435. Washington, D.C.: Nat. Acad. Sci.

Ricklefs, R. E. 1975. The evolution of cooperative breeding in birds. *Ibis* 117: 531–534.

Ricklefs, R. E. 1979. *Ecology*. New York: Chiron Press.

Ridpath, M. G. 1972. The Tasmanian Native Hen, *Tribonyx morteirii*, pts. I–III. *CSIRO Wildl. Res.* 17: 1–118.

Robertson, W. B., Jr., M. J. Robertson, and B. C. Kittleson. 1979. Social organization in a Sooty Tern (*Sterna fuscata*) colony. Abstracts, 97th Stated Meeting Amer. Ornithol. Union, College Station, Tex.

Rowley, I. 1965. The life history of the Superb Blue Wren, *Malurus cyaneus*. *Emu* 64: 251–297.

Rowley, I. 1976. Convener. Co-operative breeding in birds. *Proc. XVI Internat. Ornithol. Congr.*, pp. 655–684.

Rowley, I. 1981. The communal way of life in the Splendid Wren, *Malurus splendens*. *Zeitschrift f. Tierpsych.* 55: 228–269.

Rutter, R. J. 1969. A contribution to the biology of the Gray Jay (*Perisoreus canadensis*). *Can. Field Nat.* 83: 300–316.

Schaffer, W. M. 1974. Optimal reproductive effort in fluctuating environments. *Amer. Nat.* 108: 783–790.

Schaffer, W. M., and R. H. Tamarin. 1973. Changing reproductive rates and population cycles in lemmings and voles. *Evol.* 27: 111–124.

Schoener, T. W. 1968. Sizes of feeding territories among birds. *Ecol.* 49: 123–141.

Selander, R. K. 1964. Speciation in wrens of the genus *Campylorhynchus*. *Univ. Calif. Publ. Zool.* 74: 1–224.

Shields, W. M. 1982. *Philopatry, Inbreeding, and the Evolution of Sex.* Albany: State Univ. of New York Press.

Skutch, A. F. 1935. Helpers at the nest. *Auk* 52: 257–273.

Skutch, A. F. 1961. Helpers among birds. *Condor* 63: 198–226.

Smith, S. M. 1980a. Henpecked males: the general pattern in monogamy? *J. Field Ornithol.* 51: 55–64.

Smith, S. M. 1980b. Demand behavior: a new interpretation of courtship feeding. *Condor* 82: 291–295.

Southwood, T.R.E., 1981. Bionomic strategies and population parameters. In: May, R. M., ed. *Theoretical Ecol-*

ogy, 2d ed., pp. 30–52. Sunderland, Mass.: Sinauer.

Stacey, P. B. 1979a. Habitat saturation and communal breeding in the Acorn Woodpecker. *Anim. Behav.* 27: 1153–1166.

Stacey, P. B. 1979b. Kinship, promiscuity, and communal breeding in the Acorn Woodpecker. *Behav. Ecol. & Sociobiol.* 6: 53–66.

Stallcup, J. A., and G. E. Woolfenden. 1978. Family status and contribution to breeding by Florida Scrub Jays. *Anim. Behav.* 26: 1144–1156.

Stearns, S. C. 1976. Life-history tactics: a review of the ideas. *Quart. Rev. Biol.* 51: 3–47.

Stearns, S. C. 1977. The evolution of life history traits. *Ann. Rev. Ecol. & Syst.* 8: 145–171.

Stenger, J. 1958. Food habits and available food of Oven-birds in relation to territory size. *Auk* 75: 335–346.

Stewart, R. M., S. M. Long, and M. Stewart. 1972. Observations on the behavior of the California Scrub Jay. *California Birds* 3: 93–95.

Sutton, G. M. 1949. Meeting the west on Florida's east coast. *Florida Nat.* 22: 23–33.

Thomas, B. T. 1979. Behavior and breeding of the White-bearded Flycatcher (*Conopias inornata*). *Auk* 96: 767–775.

Vehrencamp, S. L. 1978. The adaptive significance of communal nesting in Groove-billed Anis (*Crotophaga sulcirostris*). *Behav. Ecol. & Sociobiol.* 4: 1–33.

Vehrencamp, S. L. 1979. The roles of individual, kin, and group selection in the evolution of sociality. In: Marler, P., and J. G. Vanderberg, eds. *Social Behavior and Communication*, Handbook of Beh. Neurobiol., vol. 3, pp. 351–394. New York: Plenum Press.

Verbeek, N.A.M. 1973. The exploitation system of the Yellow-billed Magpie. *Univ. Calif. Publ. Zool.* 99: 1–58.

Vries, T. de. 1975. The breeding biology of the Galapagos Hawk, *Buteo galapagoensis. Gerfaut* 65: 29–57.

Wade, M. J. 1979. The evolution of social interactions by family selection. *Amer. Nat.* 113: 399–417.

Wade, M. J. 1980a. An experimental study of kin selection. *Evol.* 34: 844–855.

Wade, M. J. 1980b. Kin selection: its components. *Science* 210: 665–669.

Watts, C. R., and A. W. Stokes. 1971. The social order of turkeys. *Sci. Amer.* 224: 112–118.

Welty, J. C. 1975. *The Life of Birds.* 2d. ed. Philadelphia: Saunders.

Westcott, P. W. 1969. Relationships among three species of jays wintering in southeastern Arizona. *Condor* 71: 353–359.

Westcott, P. W. 1970. Ecology and behavior of the Florida Scrub Jay. Ph.D. thesis, Univ. of Florida, Gainesville.

West Eberhard, M. J. 1975. The evolution of social behavior by kin selection. *Quart. Rev. Biol.* 50: 1–33.

Wilbur, H. M., D. W. Tinkle, and J. P. Collins. 1974. Environmental certainty, trophic level, and resource availability in life history evolution. *Amer. Nat.* 108: 805–817.

Williams, G. C. 1975. *Sex and Evolution.* Monographs in Population Biology, no. 8. Princeton: Princeton Univ. Press.

Wilson, E. O. 1975. *Sociobiology. The New Synthesis.* Cambridge, Mass.: Harvard Univ. Press.

Winterstein, S. R. 1980. Growth, development, care, and survival of nestling Beechey Jays. 1980. M.S. thesis, New Mexico State Univ.

Woolfenden, G. E. 1969. Breeding-bird censuses of five habitats at Archbold Biological Station. *Audubon Field Notes* 23: 732–738.

Woolfenden, G. E. 1974. Nesting and survival in a population of Florida Scrub Jays. *Living Bird* 12: 25–49.

Woolfenden, G. E. 1975. Florida Scrub Jay helpers at the nest. *Auk* 92: 1–15.

Woolfenden, G. E. 1976a. Co-operative breeding in American birds. *Proc. XVI Internat. Ornithol. Congr.*, pp. 674–684.

Woolfenden, G. E. 1976b. A case of bigamy in the Florida Scrub Jay. *Auk* 93: 443–450.

Woolfenden, G. E. 1978. Growth and survival of young Florida Scrub Jays. *Wilson Bull.* 90 1–18.

Woolfenden, G. E., and J. W. Fitzpatrick. 1977. Dominance in the Florida Scrub Jay. *Condor* 79: 1–12.

Woolfenden, G. E., and J. W. Fitzpatrick. 1978a. The inheritance of territory in group-breeding birds. *BioScience* 28: 104–108.

Woolfenden, G. E., and J. W. Fitzpatrick. 1978b. Authors' reply. *BioScience* 28: 752.

Woolfenden, G. E., and S. A. Rohwer. 1969. Breeding birds in a Florida suburb. *Bull. Florida State Mus.* 13: 1–83.

Wright, S. 1943. Isolation by distance. *Genetics* 28: 114–138.

Wright, S. 1946. Isolation by distance under diverse systems of mating. *Genetics* 31: 39–59.

Wright, S. 1951. The genetical structure of populations. *Ann. Eugenics* 15: 323–354.

Wright, S. 1978. *Variability Within and Among Natural Populations*. Evolution and the Genetics of Populations, vol. 4. Chicago: Univ. of Chicago Press.

Zahavi, A. 1974. Communal nesting by the Arabian Babbler: a case of individual selection. *Ibis* 116: 84–87.

Zahavi, A. 1976. Co-operative nesting in Eurasian birds. *Proc. XVI Internat. Ornithol. Congr.*, pp. 685–693.

Author Index

Subject Index

401

403

Library of Congress Cataloging in Publication Data

Woolfenden, Glen Everett, 1930–
  The Florida scrub jay.

  Bibliography: p.
  Includes index.
  1. Scrub jay. 2. Bird populations—Florida. 3. Birds—
Florida. I. Fitzpatrick, John W. II. Title.
QL696.P2367W66 1984   598.8′64   84-42545
ISBN 0-691-08366-5 (alk. paper)
ISBN 0-691-08367-3 (pbk.)